Salomon Borensztejn

THE BIG BANG DELUSION

Dogma or illusion ?

A dogma with feet of clay

An alternative: the temporalistic model

A probabilistic model of the Universe

CONTENTS

Preface

p 7

PART ONE

Introduction

Chapter I: Knowing about and understanding the Universe   p 12

PART TWO

Chapter II: Methodology   p 16

Proposals for ananthropic concepts

1) Speculation or critical thinking
2) Ananthropic scientific concepts
3) Critical and ananthropic analysis of multiverses

Chapter III: Anthropic and ananthropic concepts   p 24

a) The physical concept of space

b) The physical concept of time

c) The unnecesary concept of time

d) The physical concept of the speed limit, c

e) The EPR paradox (Einstein, Podolsky, Rosen – 1935)

f) The physical concept of the Law of Conservation of Energy

g) The concept of purpose

h) The concept of optimization

i) The Second Law of Thermodynamics

PART THREE

Chance, the organizer of the Universe

Chapter IV: An ananthropic probabilistic model of the Universe   p 45

Chapter V: Evidence and arguments for the ananthropic probabilistic model of the Universe   p 52

Chapter VI: Conclusions   p 60

PART FOUR

The standard Big Bang model

Chapter VII: Redshifts and the standard Big Bang model – The theoretical prediction of the Hubble constant Ho   p 65

1) The physical concept of space and expansion

2) The physical concept of time

3) The redshift z and the theoretical prediction of the Hubble constant, Ho

Chapter VIII: Consequences and weaknesses of the spatial interpretation of redshifts　　　　　　　　　　　　　　　　　　　　　　　　　　p 75

The three pillars of the standard Big Bang model:

1) Redshifts

2) The cosmic microwave background

3) Primordial nucleosynthesis

Inflationary theories

The origin of the Big Bang

The acceleration of expansion – Dark energy

Various problems: The horizon problem

The problem of flatness and critical density

The singularity problem

The problem of the homogeneous, isotropic Universe

The Hubble constant, Ho – The age of the Universe, to

Chapter IX: The evolution of galaxies – The large-scale structures of the Universe　　　　　　　　　　　　　　　　　　　　　　　　　　p 105

Chapter X: Dark matter – The Pioneer effect – The MOND theory – The Casimir effect　　　　　　　　　　　　　　　　　　　　　　　　　p 111

PART FIVE

An alternative to the Big Bang model

The temporalistic model

Chapter XI: - The concept of time and the constant To – The temporalistic hypothesis – The search for the constant To – The ratio c / G – The quantum constant G': Four quantum effects p 117

Chapter XII: Temporalistic gravitation - General relativity and temporalistic gravitation - Masses and gravitational radii p 124

PART SIX

General conclusions

Chapter XIII: Summary of critiques of the standard Big Bang model

p 139

THE BIG BANG DELUSION

PART SEVEN

Comparison between the Big Bang model and the temporalistic model

Chapter XIV: Comparison – Conclusion - Tests   p 157

PART EIGHT

Chapter XV: <u>Calculations</u>   p 190

23 PIECES OF EVIDENCE FOR THE TEMPORALISTIC MODEL
p 205

Preface

It is not enough to declare that the Big Bang is a cosmological delusion: it is necessary to provide evidence. We do not question here the honesty or competence of the countless researchers who have spent many years dedicated to the study of the standard model of the Big Bang. What we wish to underline is the significant and unacceptable way in which research in the field of cosmology has gone adrift. The absence of validation of hypotheses put forward, unfounded speculation, indifference or disregard for rigorous scientific criteria such as Popper's 'falsifiability' or 'observable facts' (*The Foundations of the Theory of General Relativity* – 1916 – Einstein) have led cosmology into a dead end. This is what we propose to show in this book.

The Big Bang, the standard model of modern cosmology, historically stems from Hubble's interpretation of the redshifts of distant galaxies. Interpreted as being caused by the recession of galaxies, these redshifts gave rise to the concepts of the expansion of the Universe, the primordial explosion and the origin of the Universe. Astrophysicists and cosmologists, experimentalists and theoreticians, have searched for evidence for this model. Redshifts, primordial nucleosynthesis and the cosmic microwave background have become the pillars of this model. Subsequently, theories of inflation attempted to make up for the serious problems faced by the standard model of the Big Bang.

In the absence of any credible alternative, the Big Bang model reigns supreme. However, as we shall show in the following chapters, the model suffers from many serious observational and theoretical inconsistencies. In fact, it is entirely based on a fragile foundation: the interpretation of redshifts as being caused by the recession of galaxies. The temporalistic model proposes a different interpretation of redshifts which does away with most of the drawbacks of the Big Bang model.

The standard model of modern cosmology, the Big Bang model, with which most researchers are in agreement, is based on a certain number of arguments and pieces of 'evidence'. Some researchers are unconvinced, and contest the validity of this model. However, at present, no competing model provides a credible alternative to the Big Bang (Fred Hoyle's steady state Universe, tired light, the symmetrical universe, Halton Arp's variable mass theory (1999), etc).

Despite its apparent robustness, the Big Bang model suffers from major weaknesses inherent both to its elaboration and to its consequences. The Big Bang's supporters naturally underline its strong points, while its critics stress its weaknesses.

In 1962, the author, dissatisfied with the Big Bang model, undertook a critical analysis of the foundations of the model which led to an alternative interpretation of the known facts which is, in his opinion, more satisfactory. This research led him to a new conception of the 'age' of the Universe and to the discovery of a parameter that he called the temporalistic constant, To, with a value of 4.5546×10^{17} seconds, or about 14.43 billion years. This constant led him to a value for the temporalistic effect or 'recession effect' at 1 Mpc of 67.71 km/s/Mpc and for Ho (the Hubble constant) of $1 / \text{To} = 1 / 4.554610 \times 10^{17}$ s. These values were established theoretically by the author in 1962.

The very latest data provided by WMAP 5 (Table 7 – Cosmological Parameter Summary – 2008) gives a value for Ho = 71.9 (+2.6 – 2.7) km/s/Mpc and for to (age of the Universe) = 13.69 (± 0.13) billion years.

Comparing the observational value and the theoretical value for H_0: 69.2 km/s/Mpc (71.9 – 2.7) for the former and 67.71 km/s/Mpc for the latter, there is a difference of 2.16%. This difference is negligible if we consider the uncertainty in the WMAP 5 data: between 3.2% (+2.6) and 3.75% (-2.7). We should add that the value of Ho provided by WMAP 5 was obtained after 80 years of research and corrections, of which 69.2 km/s/Mpc is the most recent but certainly not the final result, whereas the theoretical value proposed by the author as long ago as 1962 has not changed since then. The value for the Hubble Constant, Ho, provided by NASA is the result of a great many cosmological observations and the unremitting work of a vast number of researchers. However, due to the very nature of the observations, the accuracy of the results can only be relative (like, for instance, the distance of distant celestial bodies such as galaxies or galaxy clusters, whereas the value of the Ho constant theoretically established and proposed by the author is very accurate since it is based on the accuracy of the universal and/or quantum constants that he uses (c, G, h, e).

From Hubble's Law, v = Ho × d, where v = recession speed in km/s/Mpc, Ho = Hubble's constant in km/s/Mpc and d = distance in Mpc, we get Ho = v / d = 69.2 km/s / 3.084×10^{19} km (3.15576×10^7 s × 10^6 × $3.26 \times 2.997925 \times 10^5$ km/s) = 2.243×10^{-18} s. If the Universe has a very low matter density, which is the case, the age of the Universe, to, equals 1 / Ho = 1 / 2.243×10^{-18} s = 4.458×10^{17} s, which is around 14.12 billion years. The differences with the values obtained by the author are, as for the values of Ho, on the order of 2.15% (Ho = 67.71 km/s/Mpc and To = 4.5546 × 10^{17} s), in other words within the range of the uncertainties.

Although the value of this 'age' to of the Universe is now close to the value of the temporalistic constant To, its meaning for cosmologists is quite different.

Personal circumstances led to the author being obliged to give up his research for some 40 years, before taking it up again in 2001. Today, he proposes the temporalistic model as an alternative to the Big Bang model.

A summary, as brief as possible, of the Big Bang model is presented, with the model's concepts, assumptions, strengths and weaknesses. The lack of any alternative to the standard model of cosmology has turned it, by default, into a near dogma.

In parallel, the temporalistic model is set out, with its concepts, foundations, and its many quantum and macroscopic consequences: electric charge, fine structure constant, spectral shift, gravitation, large-scale structures of the Universe, etc, as well as tests of the theory. The author has attempted to banish any arbitrary and unverifiable speculation that is unfalsifiable, in agreement with Popper's precepts (1934). The author considers, rightly or wrongly, that a model or theory that is based on unverifiable premises and which on that basis constructs highly speculative hypotheses, cannot claim to be a valid scientific model, since it lacks the necessary rigor to be described as such.

Contrary to the claims of its supporters, the standard Big Bang model does not meet with a general consensus from researchers. Proof of this is the 'Open Letter to the Scientific Community' (cosmologystatement.org) published in *New Scientist* on 22 May 2004. This Open Letter was approved by 510 researchers and scientists from universities the world over, including many renowned figures in scientific research. The text of this letter is very critical of the standard Big Bang model. It focuses on the so-called evidence for the Big Bang (inflation, dark matter, dark energy, etc), which appears biased due to the use of adjustable parameters. Without these purely hypothetical phenomena, there would be a fatal contradiction between astrophysicists' observations and the predictions of the Big Bang theory. Moreover, according to this Letter, the Big Bang theory cannot claim any quantitative predictions validated by observations. The successes proclaimed by its supporters in fact consist in matching observations retrospectively with adjustable parameters. This model is comparable with the Ptolemaian model, which overcame drawbacks in the theory by introducing more and more layers of epicycles.

Here is an extract from the Open Letter:

"Even observations are now interpreted through this biased filter, judged right or wrong depending on whether or not they support the Big Bang. So discordant data on red shifts, lithium and helium abundances, and galaxy distribution, among other topics, are ignored or ridiculed. This reflects a growing dogmatic mindset that is alien to the spirit of free scientific inquiry".

The vehemence and deep conviction expressed by several hundred scientists fighting the dogmatic domination of the standard Big Bang model, as well as the hegemonic practices of its supporters, attest to the cosmological delusion of the Big Bang. The lack of critical thinking and the violation of centuries-old scientific principles to the benefit of dubious hypotheses, without any experimental or observational validation (unjustifiable exponential inflation, uniformity and isotropy of space contradicted by observations such as small and large-scale structures, walls, huge voids, etc) undermine the standard model of cosmology.

Although Richard Feynman said that "science is the culture of doubt", in cosmology today, doubt and divergent opinions are not tolerated, and young researchers learn to keep quiet if they have anything negative to say about the standard Big Bang model.

The standard Big Bang model can be summarized in a few lines.

13.7 billion years ago, the Universe was born from the violent explosion of a singularity. The temperature of the primordial Universe was extremely high, over 10^{32} kelvins. After 1/100 of a second, the Universe, which was extremely dense and hot (10^{11} kelvins) was made up of an undifferentiated soup of matter and light (photons, electrons and positrons, neutrinos and antineutrinos, protons and neutrons), dominated by radiation. The temperature then fell very rapidly as the Universe expanded (Weinberg, 1980).

After another 3 minutes 45 seconds, the temperature had become low enough enough to enable the formation of nuclei of deuterium, and then of helium. After 700 000 years, the temperature had fallen to a few thousand kelvins, allowing the formation of the first atoms of hydrogen and of light elements (deuterium ^{2}H, helium ^{3}He and ^{4}He, and lithium ^{7}Li).

As expansion continued, galaxies and stars appeared, followed by the large-scale structures of the Universe (galaxy clusters and superclusters, great walls, great voids, etc). The heavier chemical elements were then forged inside the stars.

Inflationary theories, elaborated in order to resolve a number of serious problems affecting the Big Bang model, are an extension of the model but independent from it.

The standard Big Bang model is based on a certain number of principles, facts, hypotheses, assumptions, consequences and interpretations, which can be enumerated and most of which we will analyze: the redshift of distant galaxies and Hubble's constant H_0; the expansion of the Universe; general relativity and the cosmological principle; theories of inflation; the cosmic microwave background; primordial nucleosynthesis and the baryon/photon ratio at the beginning of the Big Bang; the age of the Universe; the large-scale structure of the Universe; critical density and the geometrical shape of the Universe. To this list we may add the problems of dark matter, dark energy and quintessence.

A critical analysis of the standard cosmological model of the Big Bang requires a preliminary clarification of the concepts that underpin this model, like all scientific models, both in the field of cosmology and the physical sciences and in the field of the life sciences. Einstein's profound insight, "The most incomprehensible thing about the Universe is that it is comprehensible" will serve as an introduction.

PART ONE

Introduction

Chapter I

Knowing about and understanding the Universe

According to Einstein, "The most incomprehensible thing about the Universe is that it is comprehensible".

What meaning can we attribute to this insight of one of the founders of modern physics?

The fact that Nature is subject to laws that human intelligence has succeeded in deciphering appears to be incomprehensible. Over the past two or three thousand years, researchers, whether theoreticians or experimentalists, have succeeded in elaborating theories, based on facts and concepts, that explain countless physical and biological phenomena (gravitation, cosmology, biology, genetics, biological evolution, etc) and that often make it possible to make qualitative and quantitative predictions about other phenomena.

Where do these laws and theories come from, when it might be supposed that the structure of Nature and the way it works is based on chaos?

Is Nature rational, ordered and Cartesian?

Just to express this statement or assumption demonstrates, in our opinion, a perfectly anthropocentric or anthropic conception of Nature, like the geocentric conception of Ptolemaian cosmology. In order to try to understand the Universe, I propose to reject any anthropic views of Nature. Reversing the anthropic view that is so current in scientific research appears essential to me if we wish to 'understand' the Universe. This is no easy task, since it flies in the face of an age-old human attitude and a natural inclination towards 'common sense'. In the following chapter we will analyze a certain number of major scientific concepts and put forward an ananthropic interpretation (i.e. free of the bias of an anthropic interpretation).

What does 'understanding the Universe' mean?

According to scientific criteria, to understand the Universe is to find increasingly general laws that govern the phenomena of the Universe, both physical phenomena (with their mathematical expression) and biological phenomena (structures, origin and evolution of life).

a) In the physical sciences (physics, chemistry, astrophysics, gravitation, cosmology, etc), to understand is to bring together concepts and phenomena in theories, in accordance with laws and models that are usually mathematical. These concepts and theories are frequently illogical or incomprehensible, such as Newtonian gravitation (instantaneous action at a distance), Einsteinian gravitation and curvature of space-time by mass-energy (how is it possible to bend a physical space that is empty, i.e. structureless?), quantum theory and vacuum energy (contradiction in terms), the Big Bang (creation of the Universe *ex-nihilo*), etc. The irrationality of the theories and concepts is set aside in favor of their undeniable operational and predictive value.

But does this mean that we have understood the Universe?

On both the largest and smallest scales, quantum cosmology models abound: the many-worlds theory of Hugh Everett III (1957), inflationary theories (Alan Guth, 1981), the Hawking-Turok instanton (1998), the pre-Big Bang theory of Gabriele Veneziano, 1968-1991), and the ekpyrotic model (Neil Turok, Paul Steinhardt, 2001). These highly speculative models attempt to give credibility to a largely anthropocentric Big Bang. In superstring theory, the different vibrational modes of superstrings, with a length on the order of the Planck length, l_p, in a brane universe with many dimensions, some of which are hidden, make up the particles and forces which give rise to the macroscopic Universe (string theory, M-theory, supergravity theory). The 'Theory of Everything' (TOE) claims to be the ultimate explanation.

Do these different models, whether validated or not, make the Universe comprehensible? Although they may shed light on the structure and working of our physical Universe, they provide no explanation of its possible origin, or why it exists. They therefore open the way to metaphysical or religious arguments that can only be rejected by science.

b) In the biological sciences, to understand is to understand the structures, functioning, origin and evolution of living organisms.

The biological sciences study the structures of living organisms, their 'organs' and 'functions', cells, prokaryotes, archaea and eukaryotes, tissue, single-cell and multicellular organisms, phylogenies, species, genera and phyla. Specialisation in research has led to many concepts and disciplines, such as biochemistry, cell biology, molecular biology, genetics, genomics, proteomics, integrative biology, etc.

The biological sciences are marked by essentially finalistic thinking. Concepts such as 'organ', 'function', '(selective) advantage' and 'natural selection' are concepts that imply purpose and value judgments. To speak of the 'functions' of a tissue or set of tisues (organ) rather than of their 'properties' implies a finalistic way of thinking.

c) This anthropomorphic attitude cannot enable us to 'understand' the Universe.

"Basically, the anthropic principle says that we see the Universe the way it is, at least in part, because we exist" (*The Universe in a Nutshell*, Stephen Hawking, 2001). The strong anthropic principle justifies the existence of the Universe by the existence of humans. Causality is reversed. Humans become the cause, and the Universe the effect. Let us pursue this anthropic reasoning. Humans are primates. The origin of primates, several tens of millions of years ago would thus justify a 'primatopic principle', well before the appearance of *Homo sapiens sapiens*!

d) <u>Conclusion</u>: Whether in the physical or biological sciences, an anthropocentric (anthropic) viewpoint with regard to phenomena—purpose in biology; origin, end, and reason for the existence of the Universe in cosmology, etc—does not enable us to 'understand' the Universe. Anthropocentrism, in other words anthropic bias, is just about as useful for 'understanding' the Universe as geocentrism was for understanding the Solar System.

My analysis of the anthropic bias of many concepts in contemporary science runs counter to current ways of thinking. Critical thinking, which is the foundation of knowledge, is today giving way to unbridled speculation that arrogantly ignores criteria judged to be obsolete, such as 'the observable facts' (Einstein) and 'falsifiability' (Popper). It is therefore likely that this analysis will be rejected by most researchers. In no way does this take away from its credibility. The history of science teaches us that this is the usual fate for ideas that go against the reigning consensus. In any event, it is the future, however distant, of humans that will be the true judge in this matter.

The majority of the facts concerning the Big Bang model cannot be disputed: redshift of distant galaxies, cosmic microwave background, large-scale structures of the Universe (galaxy superclusters, walls, great voids), etc. We will see that the weaknesses in the Big Bang theory are essentially due to the interpretations of these phenomena, which, in our opinion, are arbitrary and unfounded. We put forward other interpretations, together with their validation of course, and an alternative model, the temporalistic model. As for the many speculative theories, which are for the most part unfounded, non-validated and impossible to validate, such as parallel universes, wormholes, the pre-Big Bang, instantons, inflationary theories, ex-nihilo creation of space and time, singularities, etc, these are anthropic concepts that must be rejected in a strictly scientific approach. They violate the criteria of observable facts (Einstein), falsifiability (Popper) and ananthropic concepts.

Current cosmology is largely based on a tendency to confuse facts and their interpretation. For instance, the redshift of distant galaxies is considered to result from the expansion of space. And yet, in reality, we only observe a fact: redshift. In no way do we observe the expansion of space. This is only an interpretation, in other words a hypothesis. This hypothesis, incidentally, is based on the concept of space, which is a muddled concept. Are we talking about a mathematical space? In that case, it cannot be invalidated physically. It simply has to be consistent. But in that case, its physical interpretation has no validity. Or are we talking about an empty physical space? In that case, it is a contradictory, i.e. anthropic concept: how can a physically empty space, in other words one without any properties, be expanding or curved? Many other concepts in the Big Bang model do not stand up to a rigorous critical analysis. The origin of the Big Bang and the resulting singularity have no known cause. These are only hypotheses, not without serious flaws, such as the concept of a singularity. Such hypotheses result from hypothetical and questionable interpretations of established facts such as the redshift of distant galaxies. Many other interpretations are possible and indeed exist, but they are sidelined by the all-powerful supporters of the Big Bang model.

PART TWO

Chapter II Methodology

Proposals for ananthropic concepts

<http://site.voila.fr/probability>

1) Speculation or critical thinking

Our first proposal is in contradiction with the spirit of our times. Imagination is a quality that researchers need, leading them off the beaten path and enabling them to put forward new solutions to new or recurrent problems (Copernicus, Galileo, Kepler, Newton, followed by Einstein; Planck; Lamarck and then Darwin; Mendel, followed by Crick and Watson, etc). However, imagination without the rigor of critical thinking can only lead knowledge into dead ends. Popper's 'falsifiability' and Einstein's 'observable facts' are today considered to be outdated concepts.

Sophisticated mathematical and computer models rule the roost. Today, the most important thing appears to be their internal consistency rather than the outdated criteria of Popper and Einstein. Contemporary astrophysics and quantum cosmology are highly representative in this respect: parallel or twin universes that we can never know about, multiverses (multiple universes), wormholes distorting time, a purely speculative exponential inflation that arbitrarily extrapolates the laws of physics in order to save the Big Bang, granular structure of space, pre-Big Bang theories, Instanton theories, information and complexity theories, etc, which make it possible to avoid the disastrous singularity of the Big Bang, *ex-nihilo* creation of matter-energy, etc. Speculation is given free reign. Experimental or observational validation has become an optional epiphenomenon. Researchers vie with each other in imagination, far from the mediocre constraints of observable reality.

We propose a measure that is today considered largely obsolete, but which continues to constitute the very foundation of scientific knowledge, namely the rejection of any speculation that is unverifiable or, put another way, that violates Popper or Einstein's precepts.

Anthropic concepts underpin all the biological sciences. Because of their historical origins, the life sciences are marked by an essentially finalistic philosophy. Concepts such as 'organ', 'function', (selective) 'advantages' and 'natural selection' are concepts that imply purpose and value judgments. Does one speak of the 'function' of electrons in an atom? No, but rather of their properties. That is a neutral assertion. To speak of the 'functions' of a tissue or set of tisues (organ) rather than of their 'properties' is to adopt a purely anthropic viewpoint. The language of biology is a finalistic and/or utilitarian one, and yet the concept of usefulness is foreign to Nature. It is improper to endow it with any kind of value judgment. Biological purpose or usefulness is therefore a purely anthropic concept, and not a scientific one.

This anthropomorphic attitude is incapable of enabling us to understand the Universe we live in. The anthropic principle, applied to the physical sciences, to Earth, planetary and space sciences and to the biological sciences, is the naive outcome of such an attitude.

"Basically, the anthropic principle says that we see the Universe the way it is, at any rate in part, because we exist" (The Universe in a Nutshell, Stephen Hawking, 2001). The strong anthropic principle justifies the existence of the Universe by that of humans. Causality is reversed. Humans become the cause, and the Universe the effect.

The anthropic principle originates in an essentially pretentious (and absurd) human viewpoint. At the current time, it is estimated that there are, in the observable Universe, around 100 billion galaxies, in other words at least 10^{22} stars with an average of 5 to 10 planets in orbit around each one of them. We can see just how laughably unimportant our tiny planet is in the Universe: just one among 10^{23} planets!

We will examine, in turn, the principal phenomena that the Big Bang model claims as evidence, i.e. the pillars on which it is based. We will see that they are not facts but rather interpretations of facts that are disputable both theoretically and factually. We will indicate the alternatives that the temporalistic model proposes, with validations to back them up.

In this book, we will attempt to avoid, as far as possible, mathematical expressions, which we will only use if absolutely necessary. Mathematical arguments for the different chapters appear in the last part of the book.

2) Ananthropic scientific concepts

We will analyze, in turn, the physical or biological concepts of space, time, maximum speed limit, the EPR paradox, the law of conservation of energy, purpose, optimization and the Second Law of Thermodynamics.

We will deliberately place ourselves within the framework of physical or biological reality. Although mathematical concepts and computer models are powerful tools for theorizing and modeling phenomena, it nevertheless remains true that mathematical models can be consistent while at the same time having no direct relationship with the physical reality of phenomena. Our study will be strictly limited to the critical analysis of these general concepts by basing ourselves essentially on their physical reality and observable facts.

How can an ananthropic concept be distinguished from an anthropic concept? Several criteria appear to us to be suitable for making this distinction:

1) An ananthropic concept must be neutral or objective with regard to Nature. In other words, it must in no way be subjective or express a value judgment. Thus, the current interpretation of quantum mechanics which closely connects the observable to the observer cannot claim an ananthropic status due to its subjectivity. Similarly, in biology, the Darwinian concepts of 'advantage' and 'natural selection', which imply a value judgment and a biological purpose should, quite correctly, be considered as anthropic concepts.

2) An irrational or speculative concept, elaborated at the expense of critical thinking, should be considered to be anthropic. This is the case for the concept of inflation and for inflationary theories in cosmology, which arbitrarily extrapolate or violate the known laws of Nature, without any observational or experimental validation. Their sole justification is the creation of ad hoc models that can be used to overcome the serious drawbacks in the Big Bang model. This is equally true for the unverifiable

concepts of the pre-Big Bang (Veneziano 1968-1991), parallel and multiple universes (Hugh Everett 1957), and imaginary instantons (Stephen Hawking - Turok 1998). Other equally speculative theories, such as the ekpyrotic theory (Neil Turok, Paul Steinhardt 2001), attempt to lend credence to a largely anthropic Big Bang model.

3) Concepts that infringe, without any real validation, what we may call the Reality Principle, in other words validated scientific facts and laws, cannot lay claim to an ananthropic status. They lack the 'falsifiability' required by this status. This is true for the concepts of instantaneous action, faster-than-light speed, *ex nihilo* creation of matter or energy, etc. They must be rejected if no observation or experiment confirms their validity.

4) Contradictory concepts naturally cannot attain an ananthropic status. For instance, the concept of a quantum vacuum <u>filled</u> with quantum fluctuations and virtual particles cannot be considered ananthropic since it is contradictory.

5) In the final analysis, an ananthropic concept is a concept that rejects the human being as a yardstick for any phenomenon, whether physical or biological (Darwinian advantage, physical optimization of motion, energy or action, etc).

3) Critical and ananthropic analysis of multiverses

Contemporary cosmology abounds with anthropic concepts, models and theories. One of the most speculative concepts is that of parallel universes or multiple universes, which go under the name of the multiverse.

The countless models of multiverse (Max Tegmark) can be classified into four principal models:

1) The simplest model stems from the application of general relativity to the Universe. According to the Big Bang theory, if we take into account the speed limit of light and the expansion of the Universe since the primordial explosion, the observable Universe is currently located 46 billion light years from the Earth. Beyond this there exist countless other universes, with physical laws similar to ours.

<u>Critique:</u>

The existence of these countless universes, beyond the observable universe, is the simple assertion of a hypothesis, without any evidence, or even the possibility of validating it. It is both totally anthropic and 'unfalsifiable'. Assuming a nearly flat Universe, within the framework of general relativity, we can simply assume that our Universe continues beyond the observable horizon without any discontinuity or any possibility of this being verified. In no way is there any question of a multiverse.

- :- :- :- :- :- :- :- :- :-

2) The theory of eternal inflation and of the 'bubble-multiverse'. This theory was developed by Andrei Linde and is based on the hypothesis of the inflation that our Universe is supposed to have undergone 10^{-35} seconds after the Big Bang. This inflationary phase lasted 10^{-32} seconds during which the Universe expanded by a factor in the region of 10^{50}, after which the Big Bang continued to evolve. Linked to string theory, the eternal inflation theory asserts that many regions of space underwent similar inflation, giving rise to an infinite number of 'bubble-universes'. The eminently speculative nature of this theory is asserted by the author.

Critique:

This entirely speculative theory is based on the concept of inflation, which is itself highly speculative (see 'Inflationary theories' – Chapter VIII) and on string theory, which at the moment, and after more than 20 years of research, has obtained no factual result (Lee Smolin - 2008), and which an impressive number of theories, between 10^{500} and 10^{1000}, are capable of explaining. This inflationary theory is the best example of the Ptolemaian system, which piles up hypotheses on hypotheses. It amounts to a strictly 'anthropic' concept that no rigorous scientific mind can accept.

:- :- :- :- :- :- :- :- :- :

3) The quantum multiverse model or many-worlds model of Hugh Everett (1957). This model applies the principle of the superposition of quantum states on the largest scales, i.e. to the Universe. It infers from this that all possible worlds co-exist, the one that we live in and those that are parallel to it. The only difference is that we can only study and understand the world in which we live.

<u>Critique:</u>

Hugh Everett's many-worlds model is based on the principle of superposition of microscopic quantum states, generalized without any justification to the macroscopic level. Quantum physics, a science that applies to the smallest scales, has never proved nor, therefore, allowed this.

Hugh Everett uses a concept that does not exist and that is therefore entirely imaginary. Moreover, there is no way of proving or 'falsifying' his theory. Thus, in totally arbitrary fashion, our macroscopic Universe is turned into an 'unfalsifiable' multiverse. Hugh Everett's many-worlds model is also a strictly 'anthropic' model.

- :- :- :- :- :- :- :- :- :- :-

4) Lee Smolin's 'cosmological natural selection' hypothesis, which stems from the 'loop quantum gravity' theory, proposes a multiverse inspired by Darwin's natural selection. According to this model, new expanding universes are born inside black holes and are given physical laws that are almost identical. In this model, there are no singularities inside black holes, and gravity becomes repulsive there. This evolutionary 'rebound' favors the production of black holes, i.e. the reproduction of other universes. By means of this process, our Universe would give rise to 10^{18} baby universes.

<u>Critique:</u>

The 'cosmological natural selection' hypothesis does not escape the criticism leveled at the other models, namely that it is speculative. The 'cosmological natural selection' model does not explain the process by which physical laws are passed on, nor how or why gravity inside a black hole is transformed into expansion. The corpus of 'cosmological natural selection' and of the creation of multiverses suffers from the same weaknesses as the other multiverse models, namely a total lack of evidence and a proliferation of assumptions.

CONCLUSION:

1) All the multiverse models are based on one or more assumptions.
2) None of the models provides evidence for its validity.
3) None of the models is 'falsifiable', in the Popperian sense.
4) None of the models puts forward 'observable facts', in the Einsteinian sense.
5) None of the models is 'ananthropic' in nature, as defined by the temporalistic model.

In summary, the various multiverse models are a good illustration of the way in which cosmology has gone adrift in the past few decades. The outcome of this is the 'dogma' of the Big Bang model. Assertions are considered, without the least justification, either as evidence, or at the very least, as predominant hypotheses (for example, the cosmic microwave background, stated to be 'fossil radiation', even though it is merely an observation at the present time; or the highly speculative concept of inflation (see 'Inflationary theories' – Chapter VIII)). No real evidence is provided for the origin of the Big Bang, for the primordial explosion, which violates the laws of physics, for the *ex nihilo* creation of mass-energy, of space and time, etc.

At the present time, the rigorous criteria of classical physics (quantum field theory, general relativity) are considered by many cosmologists to be constraints. They free themselves from such constraints by accepting that the boldest and riskiest speculations are permitted. The greatest names in cosmology, physics and mathematics subscribe to these new ways of doing research: Edward Witten, Stephen Hawking, Stephen Weinberg, etc. Others, far fewer in number, oppose this approach (David Gross).

Scientific concepts that are entirely speculative and proclaimed as such (Andrei Linde) are acceptable. In this case, they are the result of human activity that is separate from science. They are metaphysical notions, recreational activities, or science fiction. They have strictly nothing in common with the search for scientific truth.

Chapter III

Anthropic and ananthropic concepts

a) The physical concept of space

Aristotelian or Newtonian space is an absolute framework within which phenomena take place. Einstein relativized this space by including it within a four dimensional universe, spacetime (made up of three spatial coordinates and one time coordinate) where events take place (On the Electrodynamics of Moving Bodies 1905). Time and space, which are intimately linked, constitute frames of reference against which physical phenomena, such as momentum, energy, speed, etc are measured. Physical laws are invariant with regard to a change in the reference frame. In fact, Special Relativity does not relativize space but rather the spatial measurements of rigid bodies located in space, according to their state of motion or rest. Similarly, it does not relativize time but rather temporal measurements of clocks at rest or in motion. Special Relativity stems from the premise that the speed of light in a vacuum is constant.

Extending his approach, Einstein geometrized the concepts of force and gravity into an optimal (geodesic) path described by a test particle in a four-dimensional space (spacetime) curved by the presence of mass-energy ((The Foundations of the Theory of General Relativity, 1916).

If space is curved by the presence of mass-energy, it is because it appears as being different from mass-energy and as a framework where events take place. Einstein's hypothesis is that the curvature of spacetime must be zero in a vacuum, which is therefore a flat space. The vacuum itself, i.e. the existence of a fieldless space, is refuted by Einstein. Objects are not *in* space but rather have a spatial extent. The concept of empty space is thus meaningless. The geometrization of gravity by General Relativity brings about a shift from physics to geometry. However, John Wheeler points out that time and space are very different in nature and are not completely identifiable with each other. A physical space that contains neither mass nor energy is thus conceived of as an empty framework. How is it possible to conceive of the physical (not geometrical) curvature of empty space, in other words, of nothingness? Nothingness, by definition, can be neither flat nor curved. In fact, in General Relativity, Einstein describes and calculates the modification, i.e. the curvature, of Euclidean paths of a test particle in a space containing mass-energy. It is not empty space that is curved; that is impossible by definition. It is the paths of bodies in empty space that are altered by the presence in this same space of mass-energy. Empty physical space, i.e. nothingness, or put another way, the container, cannot be modified by curvature without there being a contradiction in terms. We are back in the Aristotelian framework of space. This then raises the following question: how can mass-energy, without Newtonian gravity, without curvature of empty space, affect, i.e. curve, the paths of test particles?

We have attempted to provide part of the answer to this question in the temporalistic model of gravity that we put forward (see http://site.voila.fr/nobigbang Chapter 9: Temporalistic gravitation).

In quantum mechanics, the vacuum of space is not empty. Due to Heisenberg's uncertainty principle, space is a place where quantum fluctuations take place, and it is filled with virtual particles.

In the Big Bang model, distant galaxies are moving away from each other at a speed that is proportional to their distance and with a redshift whose value is given by the Hubble constant, Ho. The redshift is interpreted as being a cosmological effect, the expansion of the Universe. The expansion of the Universe is conceived of as being an expansion of space, which carries the galaxies along with it. The usual comparison is with a balloon, or a hypersphere, which expands, carrying along the objects on its surface. The origin of this expansion is attributed to various causes, such as the Big Bang, inflation, the cosmological constant, dark energy, quintessence, the instanton, etc. Space, in the Big Bang model, therefore appears to be an ambiguous concept. Is this space abstract and mathematical, or real and physical? Is it empty, in other words, nothingness? Or is it rather the curved space-time of General Relativity? The contradiction is the same as for General Relativity. How can empty physical space, in other words nothingness, be expanding?

The physical concept of space, in contemporary physics and cosmology, is contradictory. The quantum spatial vacuum is not really empty since it is filled with virtual particles. The spatial vacuum of general relativity, in the absence of mass-energy, can be considered to be empty. How can the presence of mass-energy physically curve an empty framework? An empty physical framework is neither flat nor curved. It has no spatial dimensions. General relativity is a theory that is mathematically consistent and physically validated. Although its predictive value has long been demonstrated on a daily basis, its rationality has not. This is also true for quantum mechanics. As for the expansion of the Universe, the foundation of the Big Bang model, it suffers from the same handicap with regard to rationality as general relativity does.

Given its contradictory irrationality, the quantum concept of space or of vacuum must be considered as anthropic. This is also true for the relativistic concept of space, which curves space or the vacuum rather than curving paths in this space or vacuum. There is therefore a need to search for an ananthropic conception of space, in other words one that is rational and non-contradictory, and which integrates the substantial and indisputable results of quantum mechanics and Einsteinian relativity. Such a model is possible. It is an approach to this model that the author puts forward in his temporalistic model's notion of temporalistic gravitation (http://site.voila.fr/nobigbang) – Chapter 9: Temporalistic gravitation). The temporalistic model, based on the new quantum constant To, proposes a new interpretation of redshifts and an alternative to cosmology's standard model, the standard Big Bang model.

The redshift of distant galaxies is interpreted, in the standard Big Bang model, as being a cosmological effect caused by the expansion of the Universe, or rather of its space. In accordance with its working hypothesis, the temporalistic model interprets it as a <u>quantum and temporal</u> phenomenon, and not as a <u>cosmological and spatial</u> one. According to the temporalistic model, the redshift, z, of photons traveling through space, is the result (to the exclusion of any external interaction) of the influence of the asymmetry of time, i.e. of the existence of the quantum constant, To, on the photons. It has no relationship with the concept of 'tired light'. The author proposes an alternative to the Big Bang model of the expansion of spacetime (Chapter VII – paragraph 1 – page 68).

b) <u>The physical concept of time</u>

Just as it does for space, Special Relativity relativizes the Aristotelian or Newtonian concept of absolute time. However, as for the concept of space, Special Relativity does not relativize time but rather measurements of time, i.e. the temporal measurements of clocks, according to their state of rest or motion. The relativistic working of clocks in Special Relativity stems from the same premise, namely that the speed of light in a vacuum is constant. Nevertheless, the time coordinate keeps its preferred direction, from the past to the present and future, unlike spatial coordinates. This preferred direction of time gives rise to a 'light cone', which delimits observable events in the Universe. General Relativity keeps this temporal asymmetry.

Quantum physics, which integrated special relativity into quantum electrodynamics, has hardly altered the relativistic concept of time. It did change it, in a spatial sense, in Feynman diagrams, where the direction past > future is no longer preferred over the direction future > past (particles and antiparticles). By correlating uncertainty about energy with uncertainty about time, Heisenberg's uncertainty relations do not give a specific definition of time. Although Einsteinian relativity clearly emphasizes (the light cone) the past > future arrow of time, it abolishes the notion of time for photons. For a moving clock, time slows down. For a clock traveling at the speed of light, time would slow down infinitely. A photon traveling through a vacuum at the constant speed c is, according to Einsteinian relativity, unchanging. For the photon, time disappears and it is therefore located outside time.

In some superstring theories, the Universe is made up of eleven dimensions, including seven spatial dimensions wrapped up in Calabi-Yau spaces, and four visible dimensions in spacetime. In the time dimension, the photon does not age. "At the speed of light, time ceases to flow" (Brian Greene 2000).

At first sight, the macroscopic conception of time suffers from age-old, fundamentally anti-scientific religious and metaphysical assumptions: creation (of the Universe), primary cause, final cause, origin, creator divinities, countless myths, etc. Such assumptions have, in the past, led science to entirely anthropic cosmogonic theories and to their latest version, the Big Bang, which appears *ex nihilo*, and to the many theories that attempt to make up for the difficulties raised by the initial singularity (inflation, pre-Big Bang, etc).

In the final anlysis, time, in contemporary physics, is conceived of as a fourth spatial dimension of the Universe. Past > future asymmetry is the only parameter that distinguishes the spatial dimensions from the temporal dimension. This asymmetry, refuted by Stephen W. Hawking, is asserted by Roger Penrose (1996). If asymmetry disappears from the concept of time, there is no longer anything to distinguish the temporal dimension from a spatial dimension.

A recent experiment has nonetheless confirmed the asymmetry of time in strange elementary particles (PLEAR 1998).

Many theories, mainly in quantum cosmology, speculate about the concept of time. Hawking-Turok's extremely speculative Instanton theory conceives of the Instanton as being a tiny object simultaneously containing its own gravity, matter and spacetime, which triggers an inflationary universe. Andrei Linde is highly sceptical about this theory, which he judges to be more about getting media coverage than about physics. One question remains unanswered in this theory: what is the cause of the origin of the instanton? Alan Guth's inflation hypothesis and the many pre-Big Bang theories (Gabriele Veneziano 1968 – 1991) are basically speculative and/practically unverifiable.

These various theories cannot enjoy ananthropic status: 1) they are highly speculative and violate critical thinking; 2) they infringe the Reality Principle, since they are neither 'falsifiable' or 'verifiable'.

In both Einsteinian relativity and in superstring theory, time is abolished for the photon, which is therefore located outside time. In most cosmological models, space and time disappear before the quantum wall (situated at 10^{-43} second) or before the Big Bang situated at time zero.

To assert that space and time emerge with the Big Bang or before it (pre-Big Bang) means very precisley that mass-energy is created together with space and time out of pure nothingness. Such an assertion, which is entirely unfounded and without any validation, has more to do with science fiction than with rigorous science.

The concept of time, as it appears in quantum cosmological models, Einsteinian relativity and superstring theory can be considered to be a completely anthropic concept. It infringes both the criterion of critical thinking and the Reality Principle: 1) To assert that the photon is located outside time is pure, unverified speculation, with an obvious lack of critical thinking; 2) We have no physical evidence for the exclusion of the photon from time.

The quantum and relativistic concepts of time can therefore be considered as anthropic. So there is a need to search for an ananthropic conception of time, in other words one that does not violate critical thinking, that integrates the substantial and indisputable results of quantum mechanics and Einsteinian relativity, and that is 'falsifiable'. This is what the author puts forward in his temporalistic model based on the hypothesis of the fundamental asymmetry of time: (http://site.voila.fr/nobigbang) – (Chapter 5: The concept of time).

Analyzed critically, the concept of universal time appears as an unfounded and unnecessary concept. This is what we propose to show in the following paragraphs.

c) THE UNNECESARY CONCEPT OF TIME

1 A scientific concept

The analysis of the concept of time can be carried out philosophically or scientifically. Our research carefully avoids the quagmire of metaphysics and is in accordance with strictly scientific, rigorous criteria, and with the requirements of 'falsifiability' in the Popperian sense and 'verifiable facts' as defined by Einstein.

2 The interpretations of the concept of time

The scientific concept of time can be characterized historically. It may be said that a scientific concept of time dates back to Aristotle, who defined it as a measurement of motion. Time thus appears as an absolute framework within which phenomena take place. For Newton, time is a parameter of motion described by a differential equation. It is absolute, and inseparable from mechanistic determinism.

The concept of absolute time has been demolished by thermodynamics, special relativity, general relativity and quantum physics. According to thermodynamics, a system, made up of many elements, evolves probabilistically and irreversibly towards increasingly probable states: this is the increase in entropy. The irreversibility of systems implies an arrow of time from the past towards the future. Chaos theory, a branch of thermodynamics, weakens the predictability of phenomena without altering their irreversibility. According to the theory of special relativity, time is intimately connected with space in the concept of four-dimensional spacetime. Measuring it depends on the state of motion of the frame of reference. Spacetime is linked to mass-energy in the theory of general relativity, where the measurement of time is affected by gravity and the relative motion of frames of reference. With Heisenberg's uncertainty principle, quantum physics introduces a fundamental uncertainty into the measurement of time combined with energy, but does not express an opinion about the nature of time.

3 Anthropic and ananthropic concepts

In Chapter II (Proposals for ananthropic concepts), we indicated the criteria for ananthropic concepts which, in our opinion, make it possible to define an ananthropic concept freed from the bias of using human beings as a yardstick for biological or physical phenomena. In summary, the criteria are as follows:

1. Objectivity or neutrality with regard to Nature (to the exclusion of any value judgment);
2. Critical thinking that refutes any irrational or speculative concepts;
3. Rejection of concepts that violate the reality principle, and that lack convincing facts and falsifiability;
4. Rejection of contradictory concepts, such as curved physical space (i.e. an empty framework), a (quantum) vacuum filled with virtual particles;
5. Rejection of the insignificant inhabitants of Earth as a yardstick for the analysis of physical and biological phenomena (with its value judgments, biological purpose and advantages, physical concepts of optimization, etc).

4 The contemporary concept of time

Various aspects of time illustrate contemporary concepts of time: the passage of time, the engine of time the arrow of time, the origin of time and its possible end (Big Bang and Big Crunch), causality (effect after cause), and as we saw above, the past > future asymmetry of time (refuted by Stephen W. Hawking and asserted by Penrose 1996). In parallel with concepts of time, the flow of time has been measured from the very earliest times. The oldest known device is the gnomon (a simple stake in the ground), already used in China in 2 400 BC. This was followed successively by the sundial, the water-clock, the hourglass, clocks, chronometers, quartz clocks and the atomic clock. The current definition of the second was agreed internationally in 1967: "the second is the duration of 9 192 631 770 periods of the radiation corresponding to the transition between the two hyperfine levels of the ground state of the caesium-133 atom." The time scale that results from this is TAI, or International Atomic Time. The accuracy with which the flow of time is measured today provides no information about the nature of time.

5 Ananthropic analysis of the concept of time

This analysis is, of course, based on the essential criteria of falsifiability (Popper) and verifiable facts (Einstein).

Time, defined as a framework or a container of the Universe where events or phenomena take place, is a concept that has no justification. Without natural (physical or biological) phenomena, time cannot be materialized. This concept is therefore entirely unnecessary, as was the concept of the ether before the relativistic revolution. All the measurements of time that we listed in section 4 are measurements of the duration of physical phenomena (motion of the shadow of a gnomon or of a sundial pointer, the flow of sand, oscillations of atomic clocks, etc). So we can see that the various 'measurements of time' are in fact only 'measurements of durations' of certain physical phenomena: the motion of the shadow of a gnomon or sundial pointer, the speed at which sand flows, the number of periods of radiation, etc.

We propose that since the concept of time, considered independently of natural phenomena, is unverifiable and unnecessary, it should be eliminated, just as the concept of the ether was eliminated in the theory of electromagnetism. What remain are the various physical and biological phenomena that exist: appearance and disappearance of stars and galaxies in the cosmos, birth and death of living organisms, biological evolution, stellar evolution (Hertzsprung-Russell diagram). No material or energetic event in our Universe escapes variation. According to the first law of thermodynamics, in a closed system, for any transformation there is conservation of energy. The second law of thermodynamics establishes the irreversibility of physical phenomena, resulting in increasing entropy.

We can therefore characterize natural phenomena according to the following criteria:

1. The variability of everything that exists (atoms, stars, living organisms, etc).
2. Respect of the first law of thermodynamics, i.e. conservation of energy. This principle is expressed in Lavoisier's famous phrase: "Nothing is lost, nothing is created, everything is transformed."
3. Respect of the second law of thermodynamics, which establishes the irreversibility of physical phenomena, through increasing entropy.
4. Rejection of the anthropic concept of time as something that is independent of natural phenomena, that violates the reality principle and that is not falsifiable.
5. The neutral evolution of living organisms and species, excluding any purpose or value judgment (advantages, natural selection of the fittest, adaptations, etc.).

These criteria meet the requirements of ananthropic concepts: critical thinking, falsifiability, consistency of concepts, respect of the reality principle, neutrality, rejection of the bias of the human yardstick, etc.

Conversely, the non-respect of these ananthropic criteria leads to the rejection of the Big Bang theory (and everything that goes with it: inflation, physical singularity, expansion of space, primordial nucleosynthesis, etc), which violates the first law of thermodynamics and the reality principle, with the creation of the Universe, time, space and energy *ex nihilo*, not forgetting the many other drawbacks of the theory. We put forward an alternative to this theory, with a novel, evolving concept of the photon: http://site.voila.fr/nobigbang.

The Darwinian theory of natural selection (with its value judgments, advantages, natural selection of the fittest, functions, etc) should also be considered as an anthropic theory. It is not neutral with regard to the biological phenomena in Nature. It uses finalistic concepts and value judgments (advantages, natural selection of the fittest, purpose, and functions of organs, whereas these are only 'properties'.) We put forward an alternative to the STE (Synthetic Theory of Evolution): http://site.voila.fr/dinosaurs which integrates this theory, and which is based on a major corpus of verifications concerning biological evolution (causes of mass extinctions and in particular, of the death of the dinosaurs, hominization, the probabilistic consequences of changes in PO2 levels, etc).

Natural phenomena are variable. They have limited durations that can be measured using a reference frame as a yardstick. The measurement of the duration of astronomical phenomena (ephemeris time, ET) was based on measurements of the motion of stars, leading to the definition of the sidereal second. Since 13 October 1967, the definition of the second in the international system of units (SI) is based on a primary standard of time and frequency based on the transition between the two hyperfine levels of the ground state of the caesium-133 atom. The duration of 9 162 631 770 periods of this radiation defines the second. This primary standard is obtained from universal constants such as the speed of light, the Planck constant h-bar, and the charge of the electron (quantum constants). Defined in this way, the second does not measure 'time'. It measures the duration of a natural quantum phenomenon comprising 9 162 631 770 periods. It should be noted that TAI (International Atomic Time) is affected by the relativistic effects of gravity. In Paris it loses 6 ms/year compared to Boulder, Colorado (US), located at an altitude of 2 000 m.

6 Conclusions

The analysis of the concept of time has led us to establish that:

1. The concept of time is a scientifically unnecessary concept.
2. All phenomena, both biological and physical, evolve. This evolution takes place on different timescales (millions or billions of years for the stars, often annually for living organisms, on tiny timescales for quantum particles, highly

variable timescales for nuclei of radioactive atoms). This justifies the concept of specific durations for different classes of phenomenon.
3. The evolving character and varying durations of phenomena lead to the existence of the arrow of durations, which replaces the arrow of time.
4. Whether in the fields of biology or of physics, the evolution of phenomena is characterized by the irreversibility of phenomena. In the next chapter we will see what this specificity involves.

CAUSALITY OR PROBABILITY

Let us recall the analysis of the incorrect concept of causality and of the concept of chance in Chapter II of http://site.voila.fr/probability. The probabilistic model of the Universe presented there proposes that all the (biological and physical) phenomena in the Universe are the result of a single underlying law of chance. Chance is defined as being a relative factor of probability that replaces the absolute factor of causality, which leads to the classical concept of the determinism of the Universe (Laplace 1814). The probabilistic model of the Universe provides a large number of arguments and pieces of evidence to back up this proposition (Chapter V – Evidence and arguments for the ananthropic probabilistic model of the Universe).

The laws of probability are based on the law of large numbers, which is justified by the sizeable numbers that play a role in physical and biological phenomena. Let us recall the value of Avogadro's number, 6.0221415×10^{23} (the number of entities in a mole), the average number of neurons in a human brain, 10^{11}, the approximate number of galaxies in the observable Universe, 2×10^{11}, etc. As we explained in Chapter II of the ananthropic probabilistic model of the Universe, probability theory, which is the expression of the single law of chance, dominates all the biological and physical phenomena in the Universe.

In physics, probability theory applies to the kinetic theory of gases, statistical physics, quantum physics, thermodynamics, etc. According to the second law of thermodynamics, the entropy of a closed system in disequilibrium increases. In other words, the system evolves from a given state towards increasingly probable states (Boltzmann). For Clausius, the variation in entropy measures the degree of irreversibility of the evolution of a system. For Poincaré, entropy is a probability, i.e. it obeys the laws of chance. For Maxwell, the validity of the second law of thermodynamics is of a statistical nature and based on probabilities, due to the extremely small size of particles and to their huge number. As for quantum physics, it is entirely pervaded with probabilities and indeterminism (Schrödinger's probabilistic wave equation – Heisenberg's uncertainty principle).

The single law of chance which is expressed through probability theory also applies to biological phenomena, as is shown in Chapter V (Evidence and arguments for the ananthropic probabilistic model of the Universe) by countless biological facts which concern fields such as genetics (Jean-Jacques Kupiec 2005), biochemistry and the human brain (theory of natural selection of neuronal structures – Gerald Edelman –Biology of consciousness). In both biological and physical phenomena, then, the concept of probability replaces that of causality. The various phenomena thus obey the predominant factors of probability (entropy, probabilistic biological evolution, probabilistic molecular interactions inside cells, experimental evidence in favor of a probabilistic mechanism for gene expression—Kupiec, etc).

The application of the law of chance (or probabilities) implies a fundamental character, namely the irreversibility of phenomena. Probability theory shows that phenomena inevitably evolve towards the most probable state.

TIME AND PROBABILITY

What durations and probability have in common is their irreversibility. Natural phenomena are not stable. They constantly evolve. They don't just exist: they are changing all the time. This change brings with it duration, and at the same time obeys probability. These two characteristics of natural (physical and biological) phenomena are indissolubly linked. This can be summed up as follows: every category of natural phenomena take place over a specific duration, and is irreversible and probabilistic.

The concept of universal or even local time (Einsteinian spacetime) applying to different phenomena becomes unnecessary and totally inoperative. It must be got rid of, together with the concept that flows from it, the 'age' of the Universe.

SCIENTIFIC VALIDATION

The analysis of the concepts of time and causality has led us to eliminate them and replace them by the concepts of duration and probability. All natural phenomena have durations and a probabilistic fate. Our analysis is carried out on a strictly scientific basis. It has nothing to do with a philosophical concept or argument, which is always questionable and never falsifiable. It is therefore subject to the rigorous criteria of falsifiability (Popper) and of factual evidence (Einstein).

Natural phenomena can be classified, for convenience, into biological phenomena (relating to life) and physical phenomena (relating to phenomena of energy and inanimate matter).

1) Biological phenomena

All the phenomena in the living world are constantly changing (biochemistry, molecular biology, genomics, proteomics, biological evolution of animals and plants, etc).

The STE (Synthetic Theory of Evolution), which enriches the Darwinian theory of evolution, is an anthropic theory based on value judgments and purpose (advantages, natural selection of the fittest – genes, species or individuals -- adaptations, etc). The author proposes a strictly ananthropic model of biological evolution, validated by three examples: 1) the mass extinction at the K-T boundary, with the causes of the disappearance of the dinosaurs, and the four other major mass extinctions, 2) hominization and the correlation between hominid fossil sites and sources of iodine, 3) the probabilistic effects on animal evolution of an increase in oxygen levels (PO2).

To find out more: http://site.voila.fr/dinosaurs).

In Chapter VI of the site http://site.voila.fr/probability (Evidence and arguments for the probabilistic model of the Universe), we cite a large number of biological facts or models which validate the probabilistic evolution of biological phenomena. They concern fields such as genetics, biochemistry and the human brain (theory of natural selection of neuronal structures – Gerald Edelman – Biology of consciousness; probabilistic gene expression - Jean-Jacques Kupiec 2005; Mendel's laws; genetic mutations; the Krebs cycle; countless probabilistic biological models concerning biology, the brain, the genome, biochemistry, etc.

2) Physical phenomena

Contemporary physics abounds with probabilistic theories in most fields.

Let us recall the evidence and arguments for the probabilistic model of the Universe and the major paradigms of quantum physics whose general interpretation is probabilistic (Schrödinger's probabilistic wave equation and Heisenberg's uncertainty principle); the second law of thermodynamics and increasing entropy; Boltzmann's statistical mechanics, the basis of the kinetic theory of gases, whose third fundamental hypothesis indicates that "the state of a gas in equilibrium is that which corresponds to maximum probability".

In general relativity, test particles describe geodesics in a four-dimensional space, i.e. the shortest or optimal paths through spacetime. Geodesics (or shortest distances) represent the anthropic expression of optimization, which in the final analysis corresponds to the ananthropic concept of dominant or predominant probability.

Maupertuis' principle of least action in classical physics, and Hildebrandt's similar quantum principle (principle of minimum action) also express, anthropically, the ananthropic concept of dominant probability.

In optics, Fermat's principle stems from the same anthropic interpretation of a minimum travel time or distance between two points, whereas its real ananthropic meaning is always that of dominant probability in physical phenomena.

CONCLUSIONS

The analysis of the anthropic concepts of time and causality has led us to reject them and replace them by the ananthropic concepts of durations and probabilities. These two concepts incorporate the same feature: the irreversibility of phenomena, which indissolubly links them together.

Natural phenomena take place on very different spatial and temporal scales: from 10^{-15} m for the radius of the proton to 14.43 billion light years for the distance of the observable Universe; from the Planck time h-bar 10^{-33} s to the duration of the observable Universe, 14.43 billion years.

Durations can be measured using very different standards: the quantum atomic standard with a radiation of 9 162 631 770 periods of the cesium-133 atom which defines a duration of one second, or a cosmological standard such as the luminosity of supernovae used as standard candles, which enables their distances and therefore their durations to be inferred.

There no longer exists an absolute reference point for a universal time that applies to all phenomena, but rather diverse durations depending on the phenomena, measured using appropriate standards. The concept of time, lacking scientific validation (in the sense of Popper's falsifiability or Einstein's verifiable facts), is therefore a meaningless anthropic concept and thus unnecessary. It should therefore be eliminated. The ananthropic concepts of durations and probabilities, which are more productive, replace it advantageously.

d) The physical concept of the speed limit, c

The constancy of the speed of light in a vacuum is a premise of Special Relativity. The premise that c is constant and that its value is an upper limit for the speed of transmission of physical phenomena constitutes the foundation of modern physics, whether in (special or general) relativity or in quantum mechanics.

Nevertheless, a certain number of researchers have contemplated speeds that are faster than light, or even considerably faster (Tumulka's tachyon hypothesis), or infinite, as well as actions that are propagated instantaneously. Until now, none of these hypotheses has been validated. Concepts like this which infringe the Reality principle, i.e. validated scientific facts and laws, cannot therefore claim an ananthropic status.

At the current time, only the physical concept of the upper speed limit, c, enjoys this status.

e) The EPR paradox (Einstein, Podolsky, Rosen – 1935)

The EPR paradox refers to a strange concept which contradicts deterministic classical physics.

The EPR paradox asserts that the description of quantum phenomena by quantum mechanics is inadequate (hidden variables) and leads to a contradiction with one of the three following points: 1) the impossibility for a physical signal to exceed the speed c; 2) causality; 3) locality.

In the EPR article, the authors asked: "Can the description of physical reality be considered complete?" Following a detailed argument, their answer was that it cannot. Bohr's immediate answer took up the Copenhagen interpretation. De Broglie, who created wave mechanics in 1925, had many reservations about the Copenhagen interpretation, which he considered to be a suspect 'grey area'. After 25 years under the banner of quantum mechanics, he changed his mind and proposed the concept of the 'pilot wave'. The concept did not succeed in being accepted.

Quantum (probabilistic) formalism leads to the violation of Bell inequalities through quantum predictions that show that quantum correlations cannot be understood using classical concepts.

Experiments by Alain Aspect (1981-1982) showed that quantum mechanic's predictions were valid, that there were no hidden variables, and that Einstein and his colleagues were wrong. Of the three points quoted above, the third one was chosen and the conclusion was that there is nonlocality or non-separability of two particles in a single ensemble.

This notion of an inseparable ensemble, of entangled states, of instantaneous transmission of information (and not of a physical message, which would violate Special relativity) and of nonlocality is one interpretation of Alain Aspect's experiments. Another interpretation, that we set forward here, is possible. We replace the determinism mentioned above by the fundamental indeterminism of the quantum world, corroborated by Schrödinger's probabilistic wave equation and Heisenberg's uncertainty relationships. Replacing classical causality by fundamental probabilism makes the strange concept of nonlocality or non-separability unnecessary. Such a probabilistic interpretation is totally consistent with our probabilistic model of the Universe. Moreover, the EPR paradox perfectly expresses the incompatibility of Special Relativity and General relativity, with their entirely deterministic concepts, with quantum physics and its entirely probabilistic concepts.

Entanglement appears to imply nonlocality, a strange phenomenon that leads to the possibility of physically influencing an object without being in contact with it, or without being in contact with a succession of entities that connect it to us. In 1964, the Irish physicist John Bell asked the question, "Are the nonlocalities that seem to be present in the laws of quantum mechanics apparent or real?" According to what are known as the Bell equations, Bell concluded that no formalism was mathematically possible. Consequently, the physical world is actually nonlocal. Bell showed that no local theory can reproduce all the empirical predictions of quantum mechanics and that the predictions of any local theory must obey certain mathematical relationships, the 'Bell inequalities'. Experiments on entangled states of light (Alain Aspect) show that the predictions of quantum mechanics are confirmed even in situations where this theory violates the Bell inequalities. Ultimately, the world is nonlocal. The nonlocal influence between quantum particles depends only on whether these particles are entangled or not. The type of nonlocality encountered in quantum physics seems to call upon absolute simultaneity, which totally contradicts Special Relativity. This is Schrödinger's wave function (probabilistic wave function), which is at the heart of the nonlocal effects of quantum mechanics.

A certain number of remarks can be made about the strange concept of nonlocality:

1) Many explanations of the EPR paradox have been put forward; unfortunately, they are all speculative and lacking scientific validation (faster-than-light tachyon hypothesis, temporal nonlocality –Tumulka-, a special reference frame: the center of mass of the Universe (?), which violates Special Relativity, etc.

2) The use of two antinomic theories, probabilistic and deterministic, in the same argument.

3) The EPR paradox combines two incompatible theories with contradictory premises (exact position of a particle in a deterministic theory and fundamental lack of precision in quantum physics based on Heisenberg's uncertainty principle). The use, in the same argument, of two contradictory theories, one strictly deterministic, the other just as strictly probabilistic, can only lead to a logical dead end. The result of this is the EPR paradox, with entangled quantum states and incompatible concepts such as nonlocality (quantum theory) and locality (relativistic theory).

4) The probabilistic model of the Universe refutes any deterministic interpretation.

5) The entanglement of two particles or two quantum states, whether located at a distance of one meter or one kilometer from each other, necessarily results from the entanglement, at the outset, of two particles or two quantum states, according to the probabilistic premises of quantum physics.

6) The deterministic Universe can be considered to be an anthropic concept.

7) We propose a test for the nonlocality of two particles or quantum states. If the entanglement, at a distance, of two particles or quantum states A and B is initially observed, then if in a second phase the particle or quantum state A is modified, will this modification then be observed in particle or quantum state B?

f) The physical concept of the Law of Conservation of Energy

The first law of thermodynamics sets out the Law of conservation of energy in a closed system. Can the Universe be considered as a closed system? The answer to this question is currently much debated.

Moreover, whether in the macroscopic or microscopic world, or the living or non-living world, the *ex nihilo* appearance or disappearance of mass-energy has never been observed. Living organisms turn into other living organisms (reproduction), into energy, into biological molecules, or into inanimate matter (molecules). Molecules, atoms and their components (particles of matter and gauge bosons) change into each other but never disappear. Matter turns into energy and vice versa (photons > < electrons) but does not disappear. According to the Reality Principle, it can be considered that mass-energy can neither appear or disappear.

Models which contemplate the creation of mass-energy and spacetime *ex nihilo*, either at time zero of the Big Bang or at a time before the Big Bang, violate the Reality Principle and should therefore be considered anthropic. Only the Law of conservation of energy (more precisely, of mass-energy) enjoys an ananthropic status.

g) The concept of purpose

The concept of purpose basically concerns the biological sciences. Historically, these sciences have been marked by finalistic assumptions. The terms 'organ' and 'function' themselves imply utilitarian considerations and value judgments. An organ (heart, stomach, lung, etc) has a function (circulation, nutrition, respiration, etc). In the field of physics, an electron does not have a utilitarian role in the composition of an atom. It has certain 'properties' which enable it to bond with other atoms to form molecules. This dichotomy in the description of living organisms and inanimate matter has no theoretical or scientific justification. It is likely to have originated in cosmogonies of a metaphysical or religious nature. To attribute intentions or an ethical aspect to Nature is to lack the neutrality and objectivity required of concepts.

Darwinism and its continuation, the STE (the Synthetic Theory of Evolution), with their fundamental concepts of 'advantages' and 'natural selection', endow Nature with utilitarian concepts and finalistic intentions in biological evolution which violate the essential neutrality of Nature, and are incompatible with an ananthropic status. Natural selection (of the fittest), advantages (an anthropic notion, since Nature is neutral), adaptations (the organisms which are best attuned to their environment) are, without a shadow of doubt, anthropic concepts. Darwinism must therefore be considered to be an anthropic theory.

The author proposes a model of biological evolution which incorporates the Darwinian theory, with a new interpretation that respects the ananthropic character of Nature: 'A probabilistic model of biological evolution': <http://site.voila.fr/dinosaurs>

h) The concept of optimization

Whether in the physical or biological sciences, we come across the concept of optimization or maximum efficiency in many principles and disciplines.

Newtonian attraction, a dynamic concept, is replaced, in Einsteinian gravitation, by a kinematic concept, the spacetime geodesic (the shortest path taken by a test particle in a four-dimensional space that is more or less curved by masses and energy). Quantum physics uses the concept of a minimum ground state of energy (non-excited state of an atom). In quantum mechanics, Planck's quantum of action h-bar, minimum action, is the cornerstone of all physical phenomena. Classical mechanics is dominated by Maupertuis' Principle of least action. The importance of this Principle can be seen in quantum electrodynamics, where equations of motion, in field theories, stem from a quantum principle of least action (Hildebrandt 1998). In these various fields, motion, energy and action are at a minimum, which can be expressed as a Principle of minimalization or optimization of physical phenomena.

This minimalization or optimization of physical phenomena is found in biological phenomena. Such an optimization of biological processes can be observed at the molecular level, since in the final analysis, it is the properties of biological molecules that determine the properties of organisms. Although oxygen is not essential to life (anaerobiosis), "combined with the electron-transport chain, the (Krebs) cycle thus has the maximum efficiency encountered in biology with regard to recovering oxidation energy in the form of ATP." (Schoffeniels 1984). Through glycolysis and the fermentation pathway, anaerobic cells manufacture, from glucose, 2 molecules of ATP, whereas the same reaction carried out through respiration in aerobic cells produces 32 molecules of ATP (oxidative phosphorylation in the Krebs cycle), i.e. 16 times more energy (Mason 1992, Robert J.Huskey 1998).

The concept of optimization, which attributes to Nature a tendency to efficiency or purpose in physical or biological phenomena, cannot be considered to be ananthropic. The minimalization of these phenomena is nonetheless a fact. How can this be interpreted anthropically?

We saw earlier (Chapter II) that probability theory and its application through the law of large numbers, states that events whose probability or chances of occurring are very low take place very rarely or not at all, and that, vice versa, events that take place are those whose probability or chances of occurring are high. If we take the example of tossing a coin, the chances of it landing heads up are 1/2. If a die is rolled, the probability of any one side being face upwards is 1/6.

The mathematical developments of probability theory are complex, but this is not the place to go into its details.

It can thus be seen that, in probability theory, we find the process of minimalization or optimization that, as we have shown, exists in many physical concepts of minimal motion, energy and action: geodesics in General relativity; minimum level of energy in the ground state of a non-excited atom; Planck's minimum quantum of action h-bar; Maupertuis' principle of least action; and Hildebrandt's quantum principle of least action. The same thing applies for the optimization of biological processes (Krebs cycle). The concepts of minimalization or optimization, linked to a human value judgment, must be considered to be anthropic. Biological and physical phenomena of minimal motion, energy and action should be considered to be phenomena which, since they have high mathematical chances of occurring, therefore take place. They are, therefore, probabilistic phenomena that come under the category of ananthropic concepts.

The anthropic concepts of minimalization and optimization therefore appear, in the final analysis, to be the anthropic expression of the ananthropic concept of dominant or predominant probability.

i) The Second Law of Thermodynamics

The Second Law of Thermodynamics states that, in a closed system that is not in equilibrium, entropy is not conserved. It increases and moves towards a state of equilibrium. Entropy increases, during this change, towards a state of equilibrium. For Poincaré, entropy is a probability, i.e. it obeys the laws of chance. Chance is time-oriented; in a global system, entropy is irreversible. For Schrödinger, entropy is rather synonymous with disorder and with breakdown.

The Second Law of Thermodynamics, which minimalizes, via probability, the evolution of order in a closed system that is not in equilibrium, can be considered to be a concept of optimization like those mentioned in the preceding paragraph, i.e. probabilistic and ananthropic. We will come across other minimalist principles in physics in Chapter V.

Conclusion

The use of anthropic concepts in models or theories can only lead to flawed, i.e. anthropic, arguments. Only arguments based on ananthropic concepts can lead to ananthropically valid conclusions.

PART THREE

Chance, the organizer of the Universe

Chapter IV

An ananthropic probabilistic model of the Universe

<http://site.voila.fr/probability>

The model put forward here is a rigorously scientific model and is neither non-speculative nor metaphysical. It therefore abides by Popper's 'falsifiability' criterion and the Einsteinian requirement of 'observable facts' (The Foundations of the Theory of General Relativity, 1916). It is characterized by the following propositions:

The structure of the Universe is made up of mass-energy. The standard model of particles is the model which, at the current time, best describes the Universe on small scales. Superstring theories, which are appealing but highly speculative, have not been validated at present.

Determinism is based on an excessively broad conception of causality. Natural phenomena take place when certain physical conditions are met. Nuclear reactions inside stars only get under way when, for instance, hydrogen is available and a minimum temperature threshold is reached due to gravitational contraction. The star must also have a particular mass. Life is currently thought to be able to exist only on the basis of prokaryotic, archaeal or eukaryotic cells. What are often referred to as the cause(s) of a phenomenon are only, in the final analysis, certain predominant conditions (mass of a star, presence of hydrogen, temperature, cell organization, etc), "all other things being equal".

The most absolute expression of determinism is that of Laplace (1814). According to this conception, it would be enough to have total knowledge of the state of the Universe at a given time in order to know all its past states and predict all its future states. This theoretical conception is, in practice, physically unobservable and unachievable. Moreover, it constitutes a perfectly anthropic, exponential and arbitrary extrapolation of causality. For instance, it means that it would be possible to assert that the major biological mass extinctions on Earth were predictable even before our Galaxy, the Milky Way, was formed, several billion years ago. <u>This assertion, in the Laplacian sense, has no scientific basis and formally contradicts observations and the reality of Nature.</u>

The laws of science establish connections between phenomena. They show that when certain conditions are met, certain phenomena are certain, or highly likely, to take place (see the examples mentioned above: the presence of mass causes Newtonian gravitational attraction or Einsteinian curvature of spacetime, which is the 'cause' of the triggering of nuclear reactions inside a star; the presence of oxygen in the Precambrian is the 'cause' of the appearance of aerobic organisms, etc).

We can therefore say that, with regard to phenomena, the concept of causality actually only represents the predominant influence of certain conditions, called causes (temperature, cells, mass, oxygen, etc), out of a vast number of other conditions (minimum mass of a star, genes of cells, density of matter, presence or not of a cell nucleus, etc). Laws show the way in which the predominant, i.e. probabilistic, influence of these conditions appears.

The ananthropic probabilistic model proposes rejecting the concept of determinism or causality, an excessive and inadequate concept as we have just shown, and replace it by the concept of chance, defined as a concept of probability. The concept of determinism or causality is an anthropic concept which, incidentally, is historically marked by a clearly anthropocentric or religious background (first cause, final cause, prime mover, origin, creation, etc).

Is the Universe therefore merely the result of chance?

This is the conception put forward by the probabilistic model of the Universe.

What is chance?

Chance is generally conceived of as being the absence of any law, chaos, absolute contingency, the meeting of two independent causal series (Cournot 1843) and, lastly, unpredictability.

In fact, the real nature of chance is the negation of determinism or causality, an absolute concept, and its replacement by the relative concept of probability. In a particular state of the Universe, characterized by a large number of conditions, phenomena take place when certain conditions are met (minimum temperature for the triggering of nuclear reactions inside a star; the need for the presence of oxygen for the functioning of the eukaryotic cells of metazoans). Such conditions constitute predominant but not unique factors of probability, which are interpreted within the deterministic framework as being factors of causality.

The concept of probability is sometimes defined as subjective, sometimes as objective. Here we shall only consider probability theory as a mathematical model of the chances of an 'event' taking place, and its application, the law of large numbers or Bernoulli's law (1680). Expressed in a nutshell, this law states that events whose probability or chances of occurring are very low take place very rarely or not at all, and that, vice versa, events that take place are those whose probability or chances of occurring are high (example of Emile Borel's monkeys with typewriters, Boursin, 1986). The concept of probabilities, introduced by Blaise Pascal (1654), establishes the ratio of the number of favorable cases to the number of possible cases. If we take the example of tossing a coin, the chances of it landing heads up are 1/2. If a die is rolled, the probability of any one side being face upwards is 1/6. The physical or chemical composition of the coin or die, the height, speed or duration of the throw, etc are factors or conditions that play an insignificant role in the result of the throw. Probability stands out among a set of conditions, then, as a predominant but not unique factor. Probability orders and simplifies 'events', such as the result of a throw, according to their mathematical chances, which in this case is ½ or 1/6.

In the final analysis, probability selects, out of the many parameters that condition the production of an 'event' (in the above-mentioned example, the structure, chemical composition, kinetic energy of the object, etc), a sole parameter, the number of sides of the object (2 or 6), which simplifies the phenomenon and determines the mathematical chances to which Bernoulli's law applies.

The application of the law of large numbers is amply justified, given that natural phenomena call into play gigantic numbers: there are around 200 billion galaxies in the observable Universe; the average mass of a galaxy is around 10^{42} kg; the average number of stars in a galaxy is 100-300 billion; the Avogadro constant, $N_A = 6.022 \times 10^{23}$ mol^{-1}; the number of neurons in the human brain is around 100 billion, etc.

The fact that sizeable numbers play a role in physical and biological phenomena justifies the use of mathematics in the development of knowledge, just as the concept of space gave rise to geometry. Mathematics, which originated in the everyday life of humans, subsequently developed and freed itself from its empirical origins. It was thus able to elaborate imaginary concepts (imaginary numbers, negative numbers, imaginary time, etc). Mathematical truth has no need of physical validity. Its only validity is its consistency with its premises. Non-Euclidean geometries therefore have the same validity as Euclidean geometry. Mathematics constitutes a powerful tool for research and theorization in physics, as is shown, for instance, by the development of the theory of gravitation. This is both its strength and its weakness. Its strength, because mathematics makes it possible to draw up and verify sophisticated physical theories (statistical physics, thermodynamics, gravitation, etc). And its weakness, because its consistency, which is the only criterion of its pertinence, is incapable of validating a physical theory without experimental or observational verification. As a result, concepts such as imaginary time or time reversibility, parallel universes, wormholes, singularities, instantons, etc, which are very fashionable in contemporary astrophysics and cosmology, may be mathematically consistent, but are, physically, highly speculative and almost impossible to verify experimentally.

The strength and weakness of mathematics fully applies to computer models.

The role played by such sizeable numbers in phenomena therefore justifies the use of mathematics in the physical sciences. The role of probability that we put forward in our model is apparent in a great many fields, and we shall list these in the following chapter. Given the complexity of phenomena in the Universe and the very varied scales on which they operate, the presence of probability is not always obvious, even though it underlies them. The properties of an atom, a eukaryotic cell or a star may appear to fall under apparently very different laws, given the difference between their respective dimensional scales (around 10^{-15} m; 10^{-5} m; 7×10^{8} m). In fact, these different phenomena are the result of probability theory applied to different conditions, in the same way as determinism attributed this to causality. According to the ananthropic probabilistic model of the Universe, the phenomena of the Universe, both phyical and biological, are the result of chance, defined as the field of probability theory and the theory of large numbers. Probability theory is the fundamental law of the phenomena of the Universe. This law applies to the ultimate components of mass-energy (particles or superstrings), which exist and give rise to the whole diversity of phenomena from microscopic to macroscopic scales, by the action of probabilities and the law of large numbers. Chance is the only driver, whether visible or underlying, of all the diverse phenomena of the Universe. It replaces the concepts of causality and determinism. The laws of Nature are the expression of complex systems where probability has developed. The indeterminism of Nature, far from being a factor of chaos or disorder as is generally believed, is, in the final analysis, a factor of order and structure of phenomena on all scales. The application of probability theory makes it possible to predict phenomena.

Applied to the problem of biology, probability theory sheds light especially on the correlations between different environmental factors and the corresponding biology (biochemistry, morphology, sensory tissue, etc). According to probability theory, it is the most probable 'events' that take place. We have therefore proposed, in accordance with our previous remarks, that the current make-up of living organisms is the result of the statistically most probable interaction between environmental stimuli and the specific properties (biochemical, genetic,anatomical, behavioral, etc) of living matter. Since the environment has become more complex since the Precambrian, the evolution of living organisms is also the result of the most probable interaction. (See: 'A probabilistic model of biological evolution': <http://site.voila.fr/dinosaurs>).

The ananthropic probabilistic model of the Universe proposes that the physical and biological concepts and theories of the Universe (Newtonian or Einsteinian gravitation, quantum mechanics, superstrings, cosmology, etc, composition and evolution of organisms, etc) are the phenomenological expression of the apparent or underlying probabilistic structure of the Universe.

Knowing about or understanding the Universe?

1) In contemporary science, scientists seek knowledge about the phenomena that they study, in other words their structure and operation, whether they be stars, galaxies, genomes, quarks or strings. The ultimate goal of science is to bring together all phenomena in unified theories (such as the Theory of Everything in the physical sciences). In fact, this is to know about, rather than to understand, the Universe. Put more prosaically, scientists seek the 'how' rather than the 'why' of phenomena. The reason for the existence of

phenomena (microscopic or macroscopic, physical or biological) is not considered to be part of their domain. Metaphysics and religious myths are therefore free to deploy their obscurantist and dogmatic fantasies.

2) Physical and non-mathematical space is the framework within which macroscopic and microscopic phenomena take place. Thus conceived, space is the container of mass-energy (the content). Its only characteristic is the vacuum, or nothingness. It cannot therefore be curved. The spacetime of Special Relativity concerns the spatial measurements of rigid bodies located in space, according to their state of motion or rest, but not the vacuum of space. Similarly, time in Special relativity does not concern time itself but rather temporal measurements of clocks at rest or in motion. The geodesics of General Relativity are the paths, curved by the presence of mass-energy, of test particles in an empty physical space, but it is not this empty space that is curved. Quantum space cannot be considered to be an empty physical space since it is filled with quantum fluctuations and virtual particles. The ananthropic probabilistic model of the Universe puts forward an ananthropic temporalistic model where empty physical space, the framework where phenomena take place, is filled with gravitons, the source of temporalistic gravitation with a finite range. The geodesics of General Relativity describe minimum, i.e. probabilistic, pathways, of test particles, curved by the perturbation of the temporalistic gravity field caused by the presence of mass-energy, and not the curvature of space itself. The author suggests that the Casimir effect, the Pioneer effect and dark matter are the natural consequences of the existence of the temporalistic gravity field (Chapters IX and X: Temporalistic gravitation http://site.voila.fr/nobigbang)

3) In his ananthropic temporalistic model http://site.voila.fr/nobigbang (Chapter VII: The concept of time) based on the hypothesis of the fundamental asymmetry of time, the author puts forward a concept of time based on a new interpretation of the redshift of distant galaxies. This time, $To = 1 / Ho$ (Hubble constant), was established theoretically in 1962 by the author. Its value equals 4.5546×10^{17} seconds, in other words approximately 14.43 billion years. The latest data provided by WMAP 5 (Table 7 – Cosmological Parameter Summary – 2008) gives a value for $Ho = 71.9$ (+2.6 – 2.7) km/s/Mpc (so $Ho \sim 69.2$) and to = 13.69 (± 0.13) billion years. The SDSS (Sloan Digital Sky Survey) project, which has studied the redshift of 221 414 galaxies, has not altered this estimation.

4) Special Relativity's postulate of the constant speed of light in a vacuum, c, and its maximum speed limit for the transmission of physical phenomena, cannot be infringed without physical validation.

5) Models which contemplate the creation of mass-energy and spacetime *ex nihilo*, whether at time zero of the Big Bang, or continuously (the steady state universe), or in a pre-Big Bang period, violate the Law of conservation of energy and the Reality Principle. They are not 'falsifiable' and result from the concept of the expansion of the Universe, a hypothetical interpretation of the redshift of distant galaxies which is disputed by the temporalistic model. Cosmological expansion, i.e. the expansion of space which drags the galaxies along with it, contradicts the conception of an empty physical space that can be neither curved nor *a fortiori* expanding.

6) The concept of purpose, which essentially concerns the biological sciences, establishes a dichotomy in the description of living organisms and inanimate matter, without any theoretical or scientific justification. The very concepts of 'organs', 'functions', 'advantages', 'adaptations' and 'natural selection' imply utilitarian considerations and value judgments that are incompatible with an ananthropic status. A probabilistic model of biological evolution exempt from any notion of purpose is put forward at: <http://site.voila.fr/dinosaurs>

7) The concept of optimization or maximum efficiency exists in many physical and biological sciences. This concept appears, in the final analysis, to be the anthropic expression of the ananthropic concept of dominant or predominant probability.

8) The Second law of thermodynamics which minimizes, through probability, the increase in order in a closed system that is not in equilibrium, is a concept of optimization and therefore, in the final analysis, a concept of probability.

Chapter V

Evidence and arguments for the ananthropic probabilistic model of the Universe

The application of probability theory and the law of large numbers to phenomena on the different scales of Nature is, of course, to be found in the various scientific disciplines that study them. Unsurprisingly, it is in the 'hard' sciences, such as physics, astrophysics, cosmology, etc, that indeterminism appears most clearly. However, as we shall see later, it also turns up in countless probabilistic models in a wide range of disciplines in the biological sciences, in mathematics, and far less obviously, in the humanities and social sciences.

Physical or biological phenomena in which economy or optimization processes appear, in other words where motion, energy or action are optimal or minimal, are, as we saw earlier, the concern of probability theory. Since there are high mathematical chances or probabilities of their occurring, they therefore take place.

Physics

Interpreting quantum mechanics, and more generally quantum physics, is difficult and controversial. It is based on a certain number of assumptions that are unprovable but whose validity is operational. Schrödinger's probabilistic wave equation and Heisenberg's uncertainty and indeterminism relationships are the major paradigms amongst them, together with Planck's quantum of minimum action h-bar As we pointed out in the previous chapter, the latest data provided by WMAP 5 (Table 7 – Cosmological Parameter Summary – 2008) gives a value for Ho = 71.9 (+2.6 – 2.7) km/s/Mpc and to = 13.69 (± 0.13) billion years. The SDSS (Sloan Digital Sky Survey) project, which has studied the redshift of 221 414 galaxies, has not modified this estimation.

Comparing the observational value and the theoretical value for Ho: 69.2 km/s/Mpc (71.9 – 2.7) for the former and 67.71 km/s/Mpc for the latter, there is a difference of 2.16%. This difference is negligible if we consider the uncertainty in the WMAP 5 data: between 3.2% (+2.6) and 3.75% (-2.7). We should add that the value of Ho provided by WMAP 5 was obtained after 80 years of research and corrections, of which 69.2 km/s/Mpc is the most recent but certainly not the final result, whereas the theoretical value proposed by the author as long ago as 1962, $H_o = 67.71$ km/s/Mpc, has not changed since then. The value for the Hubble Constant, H_o, provided by NASA is the result of a great many cosmological observations and the unremitting work of a vast number of researchers. However, due to the very nature of the observations, the accuracy of the results can only be relative (like, for instance, the distance of distant celestial bodies such as stars, galaxies or galaxy clusters, whereas the value of the H_o constant, theoretically established and proposed by the author, is extremely accurate since it is based on the accuracy of the universal and/or quantum constants that he uses (c, G, h, e).

From Hubble's Law, $v = H_0 \times d$, where v = recession speed in km/s, Ho = Hubble's constant in km/s/Mpc and d = distance in Mpc, we get $H_0 = v / d =$ 69.2 km/s / 3.084 × 10^{19} km (3.15576 × 10^7 s × 10^6 × 3.26 × 2.997925 × 10^5 km/s) = 2.243 × 10^{-18} s. If the Universe has a very low matter density, which is the case, the age of the Universe, to, equals 1 / Ho = 1 / 2.243 × 10^{-18} s^{-1} = 4.458 × 10^{17} s, which is around 14.12 billion years. The differences with the values obtained by the author are, as for the values of Ho, in the region of 2.15% (H_0= 67.71 km/s/Mpc and To = 4.5546 × 10^{17} s), in other words within the range of the uncertainties.

The time To of the temporalistic Universe is a time limit for the redshift of the photon (similar to the speed limit c). In no way is it the 'age' of the Universe. In the ananthropic temporalistic model, there is no absolute time. It is necessary to conceive of the relative durations of the various phenomena and systems (duration of the evolution of stars, galaxies, galaxy clusters, etc), without any defined limits. ≤ http://site.voila.fr/nobigbang> (Chapter VII: The physical concept of time).

Probabilistic concepts

1. Ho is the cornerstone of all microscopic physical phenomena. The concept of a minimum ground state of energy (non-excited state of an atom) also plays an important role.

2. The interpretation of quantum physics is probabilistic with regard to its most significant features (Schrödinger's equation –Heisenberg's uncertainty principle).

3. According to the theory of General Relativity, a test particle describes an optimum or minimum path (geodesic) in a four-dimensional space (spacetime) which is curved by the presence of mass-energy. Whether it is spacetime or the path through spacetime that is curved by mass-energy, this path is minimal. The shortest distance or geodesic in General Relativity is an anthropic optimization concept which, as we saw earlier, is in the final analysis the anthropic expression of the ananthropic concept of dominant or predominant probability.

4. Like the geodesic in General Relativity, the Second Law of Thermodynamics, which minimizes, via probability, the change in order in a closed system that is out of equilibrium, is an optimization concept, in other words a concept of predominant probability (Poincaré).

5. In Boltzmann's statistical mechanics, which is the basis of the kinetic theory of gases, the third fundamental hypothesis states that "the state of a gas at equilibrium is that which corresponds to the maximum probability".

6. Maupertuis' principle of least action, which is fundamental throughout classical physics, states that action is minimal in all physical phenomena. This principle was applied by Feynman to quantum physics. Similarly, Hildebrandt formulated a quantum principle of least action. These principles of least action are, as we saw earlier, the anthropic expression of the ananthropic concept of dominant or predominant probability.

7. In optics, according to Fermat's principle, the path taken by light to travel from one point to another is that for which the travel time is at a minimum. This can also be stated by saying that the distance between these two points is at a minimum. These minima are also the anthropic expression of the ananthropic concept of dominant or predominant probability.

Planetary science

The current rotation/orbit ratio of 3/2 of the planet Mars corresponds to a probability of stability at this resonance of 55% (A. Correia et J.Laskar, *Nature*, 2004).

Spheres: most of the stars and planets are spheres. It is known that the sphere is the geometrical shape that minimizes the surface area of an object of a given volume.

Biological sciences

As we said earlier, advanced research in the biological sciences today increasingly favor probabilistic models over strictly deterministic models:

1. Mendel's laws: according to these laws, which gave rise to modern genetics, in gametes, the two components, one of male origin and one of female origin, for each character, separate; during fertilization, the components of both origins come together at random, i.e. probabilistically, for each character.

2. Genetics: Gene mutations, the basis of biological evolution, take place probabilistically.

3. Krebs cycle: Let us recall the maximum efficiency of the Krebs cycle in the production of energy in aerobic cells: "Through glycolysis and the fermentation pathway, anaerobic cells manufacture 2 molecules of ATP from glucose, whereas the same reaction carried out through respiration in aerobic cells produces 32 molecules of ATP (oxidative phosphorylation in the Krebs cycle), i.e. 16 times as much energy" (Mason 1992, Robert J. Huskey 1998).

4. The author proposes a probabilistic model of biological evolution which incorporates the Darwinian theory, with a new interpretation that respects the ananthropic character of Nature: A probabilistic model of biological evolution http://site.voila.fr/dinosaurs. This model proposes three probabilistic examples of biological evolution: 1) five mass extinctions (with the causes of the death of the dinosaurs at the K-T boundary, 2) hominization, 3) the increase in PAL PO2.

5. Gene expression was long described as being a deterministic process. Today, this deterministic paradigm has been disproved by a number of experimental arguments in favor of a probabilistic mechanism for gene expression. Cell data has been improved by a growing number of studies carried out at the molecular level. The life of a cell appears to be based on probabilistic mechanisms caused by non-specific molecular interactions where Brownian motion plays a predominant role (Jean-Jacques Kupiec 2005 – Paldi – 2003).

6) In every field of biological research, probabilistic models are coming to the fore today, with the use of probabilistic tools such as Bayesian methods, hidden Markov chains, laws of large numbers, and the Monte-Carlo

method. Below, we give a few examples taken from the countless probabilistic models available today:

Probabilities and biology

Biological Sequence Analysis: Probabilistic Models of Proteins and Nucleic Acids (with hidden Markov Models) (Richard Durbin, Cambridge University Press 1999-07-01)
Probabilistic modeling of biological data (Pierre Baldi ICS 277B – A unified Bayesian probabilistic framework for modeling and mining biological data ...)
Statistical Methods in Bioinformatics (probability and statistics in the bioinformatics context – Warren J. Ewens, Gregory R. Grant – Springer April 20, 2001)
L'analyse des séquences biologiques par Chaînes de Markov cachées (HMM), (Bernard Prum 1999) Learning Probabilistic Relational Models Nir Friedman (with Bayesian networks BNs) (Koller and Pfeffer - Stanford University 1998)

The brain

Cerveau, chance et chaos (Henri Korn – Université de tous les savoirs 21.10.2002)
Réseaux causaux probabilistes à grande échelle : un nouveau formalisme pour la modélisation du traitement de l'information cérébrale (Vincent Labatut – Inserm u455 – 2 March 2004).
OMEGA : calcul probabiliste de modèles de l'activité électrique des neurones (Denis Talay – INRIA – 2005)
Probabilistic brain atlases (Paul Thompson – UCLA Medical Center)
A probabilistic Framework for Region-Specific Remodeling of Dendrites in Three-Dimensional Neuronal Reconstructions (Narayanan – Narayan – Chattarji – National Centre for Biological Sciences – Bangalore – India – Neural Computation 2005)

The genome

Regulation of Genome Expression (probabilistic models of genome regulatory networks – Richard A. Young – MIT)

Expression des gènes et cancer : une question de probabilité ? (Jean-Jacques Kupiec – INSERM – 2005)
Bayes Networks and Graphical Models in Molecular Biology (MIT – Boston University Biocomputing Research – Graphical models at Kevin's site at MIT : Protein Modeling (Hidden Markov Models); System Biology, Functional Genomics, Gene Expression Analysis, Protein Protein Interaction (Bayes Networks); Gene Expression (Microarray) Analysis, Networks, Pathways (Bayesian Network); Biological Data Integration (Bayesian Framework); Protein Protein Interaction and Functional Annotation (Markov Random Field Approaches); DNA Sequence Analysis (Bayes Networks); Genetics, Phylogeny Linkage Analysis (hidden Markov phylogeny)
Probabilistic Models in Computational Molecular Biology (Stanford University, Stanford, CA – 2000)
Rich probabilistic models for genomic data (Eran Segal – August 2004)
A probabilistic theory for cell differentiation (J-J Kupiec – 1986)
Probabilistic discovery of overlapping cellular processes and their regulation (Annual conference on Research in Computation Molecular Biology – Alexis Battle, Eran Segal, Daphne Koller – Stanford University, Stanford, CA)
Probabilistic models of Proteins and Nucleic Acids (HMMs – Durbin-Cambridge, Eddy-Washington University, Krogh-Lyngby-Denmark, Mitchison – 1998)
Probabilistic code for DNA recognition by proteins of the EGR family (Benos, Lapedes, Stormo - J Mol Biol. 2002 Nov 1)
Recognizing complex, asymmetric functional sites in protein structures using a bayesian scoring function (Wei, Altman – Journal of Bioinformatics and Computational Biology)
A probabilistic view of gene function (Fraser, Marcotte – Nature Genetics – 27 May 2004)
Differential Proteomics via Probabilistic Peptide Identification Scores (Colinge, Chiappe, Lagache, Moniatte, Bougueleret – Anal. Chem. 2005)
A probabilistic functional network of yeast genes (Lee, Date, Adai, Marcotte – Science 2004 Nov 26)

Chemistry - Biochemistry

Amazing cellular biochemistry in terms of molecular networks (computational approaches within Bayesian formalism – Xia, Yu, Jansen,

Seringhaus, Baxter, Greebaum, Zhao, Gerstein – Annual Review of Biochemistry – July 2004)
A Thermodynamic-Probabilistic Analysis of Diverse Homogenous Stoichiometric Chemical Reactions (Garfinckle – J. Physical Chemistry 2002)

Populations

Theory of Probability (Chance plays a major role in the dynamics of a population – Joe Romano – Biomathematics 2005)
Stochastic models for biological populations – Genealogies and spatial structures (Birkner and all. – Dutch-German Bilateral Research Group "Mathematics of random spatial models from physics and biology")

Various

John Gosline (1984) showed that it was a stochastic arrangement of amorphous chains of proteins that gives spider silk its unique properties (461 times stronger than steel).
Biased random walk (biochemistry) enables bacteria to search for food and flee from harm (Wikipedia)
Perception active des formes 3 D dans le cadre d'un modèle bayésien (Jacques Droulez –CNRS – Collège de France – 8 December 2004)
A Probabilistic Approach to Large-Scale Association Scans: A Semi-Bayesian Method to Detect Disease-Predisposing Alleles (Steven J. Schrodi - Statistical Applications in Genetics and Molecular Biology – November 1, 2005)
Understanding the LDL receptor Structure through Probabilistic Models (using HMMs of the LDL receptor – MIT Computational Biophysics Laboratory – October 2005)
A Web-Based System for Public-Private Sector Collaborative Ecosystem Management (construction of probabilistic models of ecosystems processes – Timothy C. Haas – University of Wisconsin-Milwaukee)
Probabilistic Basecalling (Speed, Li, Nelson, Cawley – University of California, Berkeley – January 1999)
A probabilistic analysis of a greedy algorithm arising from computational biology (Daniel G. Brown – Cornel University)
A probabilistic model of mosaicism based on three histological analyses of chimaeric rat liver (Iannaccone, Weinberg, Berkwits – Northwestern University Medical School, Chicago II)

Mathematics

Chance is not absent from this 'pure' branch of scientific research: "These concepts of chance and unpredictability, which are fundamental in classical physics and in quantum physics, also lie at the heart of pure mathematics." (Chaitin – La Recherche December 2004 N° 381)

"...the decimals of π do seem to appear at random: those who have looked for regularities in the distribution of the decimals have found nothing, even with very powerful statistical tests." (Simon Plouffe – La Recherche – December 2005 – N° 392)

Social sciences and humanities

Insurance: to predict trends in many kinds of data, such as interest rates, growth of GDP, trends in the fertility rate, etc, or to establish premiums according to different risks (fire, life, various), actuaries apply probability theory.

The increasing use of statistics in many fields is evidence of the importance researchers attach to probability theory and to the law of large numbers.

Chapter VI

Conclusions

Summary

Our research has led us to reject causality and its absolute determinism, which do not express the reality of phenomena. The latter do not have single defined causes, but take place when a certain number of conditions are met (nuclear reactions inside a star when a certain temperature and mass are reached, nuclear genes in eukaryotic cells, etc). Laws show the way in which the predominant, i.e. probabilistic, influence of certain conditions appears, among a host of other conditions.

If "the most incomprehensible thing about the Universe is that it is comprehensible" (Einstein), this is essentially because we use the human being as a yardstick for our knowledge, which necessarily leads to anthropic concepts. We have attempted, through a critical analysis of various concepts (space, time, the speed limit c, the law of conservation of energy, purpose, optimization and the Second Law of Thermodynamics) to show the anthropic nature of many of these concepts. We have proposed to grant some of these concepts an ananthropic status (space, time, optimization, purpose, etc). The ananthropic status of a concept is its independence with regard to the human yardstick. It must meet a certain number of rigorous criteria: it must be neutral and objective with regard to Nature, be devoid of a utilitarian or finalistic bias, and without value judgement (Darwinian advantage); it must respect critical thinking and exclude unverifiable speculations (pre-Big Bang, parallel universes); it must not violate the Reality principle and must respect Popperian 'falsifiability' and Einstein's 'observable facts' (faster-than-light speed, instantaneous action); it must not be contradictory or muddled (quantum vacuum or nothingness filled with virtual particles; Einsteinian spacetime or nothingness curved by mass-energy).

The rejection of causality and its absolute determinism, and the critical analysis of anthropic concepts has led us to propose an ananthropic probabilistic model of the Universe. This model proposes a new approach to gravitation, where empty physical space, the framework where phenomena take place, is filled with a field of gravitons, the source of temporalistic gravitation with a finite range. <http://site.voila.fr/nobigbang> Chapter XII: Temporalistic gravitation).

The first consequence of the ananthropic probabilistic model of the Universe is the refutation of the expansion of the Universe and of everything derived from it, such as the Big Bang, inflation, birth of the Universe, etc).

In the field of biological sciences, the very concepts of 'organs', 'functions', 'advantages', 'adaptations' and 'natural selection' imply utilitarian considerations and value judgments that are incompatible with an ananthropic status. The ananthropic probabilistic model of the Universe proposes a probabilistic model of biological evolution which incorporates Darwinian theory, with a new interpretation of natural selection that respects the ananthropic character of Nature: A probabilistic model of biological evolution http://site.voila.fr/dinosaurs.

The ananthropic probabilistic model of the Universe proposes a certain number of pieces of evidence and arguments (Chapter V):

In the physical sciences: quantum physics is dominated by probabilism (Schrödinger's probabilistic wave equation, Heisenberg's uncertainty relationships); the Second Law of Thermodynamics; Boltzmann's statistical mechanics, the basis of the kinetic theory of gases, where the third fundamental hypothesis states that "the state of a gas in equilibrium is that which corresponds to maximum probability"; the ananthropic interpretation of geodesics in General Relativity, of Fermat's principle, of Maupertuis' classical and Hildebrandt's quantum Principle of least action.

In planetary science, the current rotation/orbit ratio of 3/2 of the planet Mars corresponds to a probability of stability at this resonance of 55%; most stars and planets are spheres, and we know that the sphere is the geometrical shape which minimizes the surface of an object of a given volume.

In the biological sciences, Mendel's laws are probabilistic laws; in genetics, gene mutations, the basis of biological evolution take place probabilistically; the Krebs cycle is a process of optimization, i.e. a probabilistic one. The life of a cell is likely to be based on probabilistic mechanisms due to non-specific molecular interactions where Brownian randomness plays a predominant role. The probabilistic model of biological evolution', proposed by the author: 'A probabilistic model of biological evolution' http://site.voila.fr/dinosaurs proposes three probabilistic examples of biological evolution: 1) five mass extinctions (with the causes of the death of the dinosaurs at the K-T boundary, 2) hominization, 3) the increase in PO2 PAL.

In the humanities and social sciences, all insurance schemes are based on the application of probability theory by actuaries.

The probabilistic model of the Universe is not a simplistic model. It simply replaces causality and absolute determinism by chance, defined as probability. The presence of chance in the phenomena of the Universe gives rise to their complexity. Events whose probability or chances of occurring are very low take place very rarely or not at all, and, vice versa, events that take place are those whose probability or chances of occurring are high. The 'laws' of the Universe are therefore merely the expression of chance, applied to the complex structure of Nature's basic building bricks (quarks, gluons, etc, or superstrings), moving within the framework of a space filled with gravitons. Phenomena are events, since they evolve (redshift of the photon) without absolute time but with various relative durations.

We can now attempt to answer Einstein's insight: "The most incomprehensible thing about the Universe is that it is comprehensible".

If we wish to attempt to understand the Universe, it is necessary to adopt an ananthropic attitude. What is the position of Earth's inhabitants in the observable Universe today? The number of planets can be estimated at around 1×10^{23} to 2×10^{23} = 200 billion galaxies containing an average of 200 billion stars with approximately 5 to 10 planets per star. In other words, the planet Earth represents $1 / 10^{23}$ of the observable planets. And the human brain, with its weaknesses and above all its prejudices, is supposed to be a suitable yardstick with which to decipher the Universe where humans live! How pretentious and how absurd! This is anthropocentrism to the power 10^{23}!

In Chapter III, we analyzed various anthropic and ananthropic concepts. We saw that, in order to 'understand' the Universe, it is necessary to rid anthropic concepts of their bias. Ananthropic concepts alone are capable of helping us to 'understand' rather than 'know about' the Universe. The need for concepts to be ananthropic in nature is comparable to the need for objectivity in science.

Ananthropic concepts enable us to find answers to age-old anthropic questions that have lastingly been obscured by myths and religions.

Why does the world exist?
When was the world created?
Will the world come to an end? If so, when?
Where do we come from? Who are we? Where are we going?

These existential questions have often been received more or less sophisticated answers: metaphysical, religious, literary or poetic, and even pseudo-scientific ones.

In fact, these questions have no answers, because they have no ananthropic meaning. They are merely anthropic questions, and as such are flawed.

Why does the world exist? The existence of the world is a fact. To give it a meaning is an anthropic attitude, one that gives Nature a purpose. This question has no ananthropic validity. It is therefore meaningless and can have no answer. This question is just as absurd and devoid of any possible rigorous answer as, for instance, the following question: are apples happy to be green?

When was the world created? The world was not created. The existence of the world is a fact. Until now, science has never observed any *ex nihilo* creation of matter, or energy, or of anything at all. For what illogical or speculative reason should the Universe be an exception to this? The question of the creation of the Universe is an essentially religious and anthropic concept. This question has no ananthropic validity. It is therefore meaningless and can have no answer.

Will the world come to an end? And when? The world will not come to an end. The reasoning is identical to that regarding the creation of the Universe.

Where do we come from? Who are we? Where are we going? The existential questions that humans ask lie derive from a purely anthropic emotional attitude. Nature, and humans are a part of Nature, has no meaning or purpose. Such questions have no ananthropic validity. They are therefore meaningless and can have no answer.

We propose this answer to Einstein's insight, "the most incomprehensible thing about the Universe is that it is comprehensible".

The Universe is not governed by causal, independent, rational 'laws' that can therefore be understood by human intelligence. It obeys a single law, the law of chance, applied to the components of the Universe. However, as we pointed out in Chapter II, far from being a factor of disorder and chaos, as is usually believed, chance, defined as the domain where probability theory and the law of large numbers apply, is a factor of order and of predictability of phenomena. This is what Einstein describes as being 'comprehensible'.

Our analysis of the anthropic bias of many concepts in contemporary science runs counter to current ways of thinking. In any event, it is the future, however distant, that will be the true judge in this matter.

We shall examine one by one the 'evidence' for the standard Big Bang model, and firstly the three 'pillars' that are supposed to establish the validity of the model: 1) redshifts; 2) the cosmic microwave background (CMB or CMBR); 3) primordial nucleosynthesis. We will then carry out a critical analysis of the many concepts and problems relating to the Big Bang model: expansion and the accelerating expansion of space, the problems of the homogeneity and flatness of the Universe, inflationary theories, the age of the Universe, the large-scale structures of the Universe (filaments, galaxy clusters and superclusters, great walls, great voids), the singularity, dark matter, dark energy, magnetic monopoles, the cosmological constant, etc.

We will carefully distinguish the facts and their interpretation. As we shall see, there is very frequently a tendency to confuse facts and their interpretation. The first example concerns redshifts. Researchers have never observed a phenomenon of recession of galaxies or expansion of space. They have observed, and still do so, the redshifts of distant galaxies. These are indisputable facts. But then these facts are interpreted as recession of galaxies and expansion of spacetime in the standard model of the Big Bang. These interpretations can be disputed, and indeed are, by other theories. Many other interpretations have been proposed: gravitational effects, the Wolf effect between two separated sources, gaseous matter in space (Marmet model 1989), Symmetric Theory (1997), variable mass theory (Halton Arp 1999), plasma universe (Hannes Alfvén 1989), etc. These various interpretations have not been accepted by the research community. For want of an alternative, the standard Big Bang model has prevailed. A great many scientists have accepted it by default. Which in no way means that it is correct.

PART FOUR

The standard Big Bang model

See calculations: CHAPTER XV page 181

Chapter VII

Redshifts and the standard Big Bang model

The theoretical prediction of the Hubble constant, Ho

Redshifts and the expansion of the Universe – The standard Big Bang model – The concept of time

The standard Big Bang model

In the 1920s, Hubble discovered that, beyond the Milky Way, the galaxies appeared to be moving away from us at a radial velocity proportional to their distance. He inferred this from the observation of the redshifts of distant galaxies, attributed to the Doppler effect. The Hubble constant, Ho, measured this recession of the galaxies according to the law v (speed in km/s) = Ho (in km/s/Mpc) × d (distance in Mpc – millions of parsecs).

In fact, the Hubble constant is really a parameter, since its value can vary.

A considerable number of redshifts are known today. NASA's IPAC database listed 153 000 (2001). The SDSS (Sloan Digital Sky Survey) project has already studied the redshift of over 221 414 galaxies.

The term 'radial velocity' is preferably used for Doppler motion, while 'speeds' are reserved for cosmological effects. At present, the redshifts of distant galaxies are considered to be cosmological phenomena caused by the expansion of the Universe.

Beyond the nearby universe, redshifts are dominated by cosmological expansion. In the Friedmann-Lemaître model, the mathematical description of cosmological expansion, distances are defined in the terms of the Robertson-Walker metric, which is the most general mathematical description for a uniform, homogeneous space that is contracting or expanding.

The standard Big Bang model results from the equations of general relativity, the cosmological principle, and Einstein's hypothesis of a homogeneous and isotropic Universe. The best measurements of the expanding Universe are at present the distances of Cepheid variable stars, the Tully-Fisher relation between the speed of rotation of a spiral galaxy and its luminosity, and Type 1a supernovae (WMAP 5 2006 - 2008).

The density of the Universe determines its geometrical shape and its fate. To obtain a stable Universe, Einstein proposed a cosmological constant or energy density of the vacuum. When Hubble showed that the Universe was expanding, Einstein rejected his cosmological constant, declaring that it was the biggest blunder of his life.

At present, it is thought that expansion slowed down after the Big Bang, and that it then speeded up around six billion years ago.

Critiques:

The expansion of the Universe and the recession of galaxies are not observational data. They stem from an interpretation of the redshifts of distant galaxies. Many other interpretations have been proposed, but not accepted.

The author's model, the temporalistic model, based on the quantum constant To, proposes a new interpretation of redshifts and an alternative to the Big Bang model.

The redshift of distant galaxies is thus interpreted, in the standard Big Bang model, as being a cosmological effect caused by the expansion of the Universe. In accordance with its working hypothesis, the temporalistic model interprets it as a quantum and temporal phenomenon, and not as a cosmological and spatial one. According to the temporalistic model, the redshift, z, of photons traveling through space, is the result (aside from any external interaction) of the influence of the asymmetry of time and of the existence of the temporalistic constant, T_0, on the parameters of photons. This has no relationship with the concept of 'tired light'.

The interpretation of redshifts as a consequence of the expansion of the Universe and of the recession of galaxies is based on the inconsistent, and therefore anthropic, concepts of space and time that we examined in Chapter III. Let us recall the fundamental elements of our previous analysis:

The physical concept of space and expansion

In the Big Bang model, distant galaxies are moving away from each other at a speed that is proportional to their distance and with a redshift whose value depends on the Hubble constant, H_0. The redshift is interpreted as being a cosmological effect caused by the expansion of the Universe. This expansion of the Universe is conceived of as an expansion of space which drags the galaxies along with it (the usual comparison is with that of a balloon or hypersphere that expands, pulling along objects on its surface). The origin of the expansion is attributed to various causes, the Big Bang (the primordial explosion), inflation, the cosmological constant, dark energy, quintessence, the instanton, etc. Space, in the Big Bang theory, therefore appears to be an ambiguous concept. Is this space abstract and mathematical, or real and physical? Is it empty, in other words, nothingness? Or is it rather the spacetime of General Relativity? In General Relativity, according to Einstein's vacuum equations, Einstein's hypothesis is that the curvature of spacetime is zero in a vacuum, which is therefore a flat space. Empty space, in other words "a space without a field", is refuted by Einstein. In the preface to the 9th edition of his book 'Relativity – The Special and the General Theory', he notes: "Objects are not located in space but rather have a spatial extent. Seen this way, the concept of empty space loses its meaning." General relativity, with its concept of geometrization of physics, thus gets rid of the physical concept of space and of the vacuum.

The discovery of the redshift of the light emitted by distant galaxies (Vesto Slipher and Edwin Hubble), previously predicted by Georges Lemaître, is interpreted as being an expansion of the Universe and as a confirmation of the theory of General Relativity.

The FLRW (Friedman – Lemaître – Robertson – Walker) metric is used to describe the Universe. The FLRW model was solely used as a first approximation because of the simplicity it brings to calculations. Models that take into account density fluctuations were then added to the FLRW model. In a strictly FLRW model, there are no galaxy clusters, , no stars, no planets, and no biological organisms, because these objects are far denser than the mean density of the Universe.

Most cosmologists agree that the part of the Universe that is observable is well approximated by a nearly FLRW model, in other words a model that follows the FLRW metric apart from primordial density fluctuations. The theoretical implications of these various expansions appear to be well understood, and the current goal is to make them consistent with the observations carried out by the COBE and WMAP satellites.

Oddly enough, it must be said that Einstein never agreed with the interpretation of an expanding Universe, and for many years he tried in vain to put forward an alternative interpretation, 'tired light'. On the other hand, John Wheeler pointed out that time and space are very different in nature and are not completely identifiable with each other.

Operationally, General Relativity, despite certain mathematical difficulties, is a success. However, on the theoretical level, it can be seen that its creator was permanently dissatisfied with some of its consequences (especially the expansion of the Universe).

The physical concept of space, in contemporary physics and cosmology, turns out to be ambiguous and contradictory. The quantum spatial vacuum is not empty since it is filled with virtual particles. The spacetime of general relativity, in the absence of mass-energy, can be considered to be empty. However, this vacuum i.e. "a space without a field", is refuted by Einstein. How can the presence of mass-energy physically curve an empty framework? An empty physical framework is neither flat nor curved. It has no spatial dimensions. General relativity is a theory that is mathematically consistent and physically validated. Its predictive value has long been demonstrated daily. But its rationality has not. This is also true for quantum mechanics. As for the expansion of the Universe, the foundation of the Big Bang model, it suffers from the same handicap with regard to rationality as general relativity.

Given its contradictory irrationality, the quantum concept of space or of vacuum must be considered as anthropic. This is also true for the relativistic concept of space, which curves space or the vacuum rather than curving paths <u>within</u> this space or vacuum. There is therefore a need to search for an ananthropic conception of space, in other words one that is rational and non-contradictory, and which integrates the considerable and indisputable results of quantum mechanics and Einsteinian relativity. Such a model is possible. It is an approximation to this model that the author puts forward in his temporalistic model based on the existence of a new quantum constant, To (<<u>http://site.voila.fr/nobigbang</u>>). The temporalistic model, based on the quantum constant To, proposes a new interpretation of redshifts and an alternative to the standard model of cosmology, the standard Big Bang model.

As we said earlier, this interpretation considers redshift to be the result (aside from any external interaction) of the effect of the asymmetry of time, i.e. of the effect of the existence of the quantum constant To, on photons. It has no relationship with the concept of 'tired light'.

The physical concept of time.

Just as it does for space, Special Relativity relativizes the Aristotelian or Newtonian concept of absolute time. However, as for the concept of space, Special Relativity does not relativize time but rather measurements of time, i.e. the temporal measurements of clocks, according to their state of rest or motion. The relativistic working of clocks in Special Relativity stems from the same premise, namely that the speed of light in a vacuum is constant. Nevertheless, the time coordinate keeps its preferred direction, from the past to the present and the future, unlike spatial coordinates. This preferred direction of time gives rise to a 'light cone', which delimits observable events in the Universe. General Relativity keeps this temporal asymmetry.

At first sight; the macroscopic conception of time suffers from age-old fundamentally anti-scientific religious and metaphysical assumptions: creation (of the Universe), primary cause, final cause, origin, creator divinities, countless myths, etc. Such assumptions have, in the past, led science to entirely anthropic cosmogonic theories and to their latest version, the Big Bang, which appears *ex nihilo*, and to the many theories that attempt to make up for the difficulties raised by the initial singularity (inflation, pre-Big Bang, etc).

Quantum physics, which incorporated special relativity into quantum electrodynamics, has hardly altered the relativistic concept of time. It did change it, in a spatial sense, in Feynman diagrams, where the direction past > future is no longer preferred over the direction future > past (particles and antiparticles). By correlating uncertainty about energy with uncertainty about time, Heisenberg's uncertainty relations do not give a specific definition of time either. Although Einsteinian relativity emphasizes (with the light cone) the arrow of time past > future, it abolishes the notion of time for photons. For a moving clock, time slows down. For a clock traveling at the speed of light, time would slow down infinitely. A photon traveling through a vacuum at the constant speed c is, according to Einsteinian relativity, unchanging. For the photon, time disappears and it is therefore located outside time.

In superstring theories, the Universe is made up of eleven dimensions, including seven spatial dimensions wrapped up in Calabi-Yau spaces, and four visible dimensions in spacetime. In the time dimension, the photon does not age. "At the speed of light, time ceases to flow" (Brian Greene 2000).

In the final anlysis, time, in contemporary physics, is conceived of as a fourth spatial dimension of the Universe. Past > future asymmetry is the only parameter that distinguishes the spatial dimensions from the temporal dimension. This asymmetry, refuted by Stephen W. Hawking, is asserted by Roger Penrose (1996). If asymmetry disappears from the concept of time, there is no longer anything to distinguish the temporal dimension from a spatial dimension.

A recent experiment has nonetheless confirmed the asymmetry of time in strange elementary particles (PLEAR 1998).

Many theories, mainly in quantum cosmology, speculate about the concept of time. Hawking-Turok's extremely speculative Instanton theory conceives of the Instanton as being a tiny object simultaneously containing its own gravity, matter and spacetime, and triggering an inflationary universe. Andrei Linde is highly sceptical about this theory, which he judges to be more about getting media coverage than about physics. One question remains unanswered in this theory: what is the cause of the origin of the Instanton? Alan Guth's inflationary hypothesis and the many pre-Big Bang theories (Gabriele Veneziano 1968 – 1991) are basically speculative and/or practically unverifiable.

In most cosmological models, space and time disappear before the quantum wall (situated at 10^{-43} second) or before the Big Bang situated at time zero.

The quantum or relativistic concept of time can therefore be considered as anthropic. There is therefore a need to search for an ananthropic conception of time, in other words one that does not violate critical thinking, which integrates the considerable and indisputable results of quantum mechanics and Einsteinian relativity, and which is 'falsifiable'. This is what the author puts forward in his temporalistic model based on the hypothesis of the fundamental asymmetry of time: (http://site.voila.fr/nobigbang) – Chapter 5: The concept of time).

The redshift z and the theoretical prediction of the Hubble constant, Ho

The redshift z of distant galaxies is interpreted, in the standard model of cosmology, as being a cosmological effect caused by the expansion of the Universe. The temporalistic model interprets it as a quantum and temporal phenomenon, and not as a cosmological and spatial one. The redshift, z, of photons traveling through space, is the result (aside from any external interaction) of the influence of the asymmetry of time, in other words of the existence of the temporalistic constant, To, on the parameters of photons. It has no relationship with the concept of 'tired light'.

When a photon is emitted by a distant light source, such as an optical electron in an atom in a star, it travels through space. Its energy is characterized by its vibrational frequency: $E = h w$ (where E = energy, h = the Planck constant, and w = frequency). According to the temporalistic model, and unlike the classical expression, the existence of the constant To means that this energy is not stable. It changes, together with the parameters that are related to it. If the travel time is t, the loss of energy ΔE

will be such that E - E' / E = t / To (where E is the energy emitted and E' the energy received). If the photon loses energy (as in the Compton effect), the changes in the wavelength can be calculated: $z = \lambda' - \lambda / \lambda = v / c = t / To$ 'where z is the redshift, λ the emitted wavelength, λ' the observed wavelength, v the apparent speed of the galaxy, and c the speed of light).

Let us examine the shift in the wavelength z of the photon caused by the existence of the constant To and as it appears in the redshift of distant galaxies. A photon, emitted in a distant galaxy at time Te, travels through space and reaches the observer at time Tr, in the reference frame of the observer. This photon travels during a time Tr - Te = distance of the galaxy / c = t. The shift in the wavelength of the photon, z, equals t / To. We can immediately see that this expression of the shift in wavelength is similar to the expression of the radial cosmological effect z = vr/c (where vr is the radial velocity). In the temporalistic model, the variation in the wavelength of a photon is proportional to the ratio between the travel time of the photon and the constant To. In the radial cosmological effect, the variation in the wavelength of a photon is proportional to the ratio between the radial velocity of the light source (compared to the observer) and the speed of light. In both cases, this is a relationship between a parameter (duration or speed) and the limiting physical constant for these parameters, To or c. However, the physical significance of the two cases is very different. In the cosmological effect, the light source is moving and the wavelength of the photon, within the photon's frame of reference, does not vary. The cosmological effect is a spatial effect. In the temporalistic model, the increasing wavelength of the photon is a temporalistic effect caused by the existence of the constant To. The light source is staionary and, within the photon's reference frame, the wavelength of the photon increases. The lengthening of the vibration of the photon, i.e. of its wavelength, as it travels through space, is a temporal or temporalistic effect. This temporalistic 'reddening' is considered, in the standard Big Bang model, to be a cosmological effect of a spatial nature, and is interpreted as a recession of galaxies, giving an apparent retrograde velocity, or 'recession effect', to distant galaxies.

According to the temporalistic model, as soon as the photon is emitted, the existence of the temporalistic constant To becomes apparent by a reddening (redshift) of its wavelength without any external intervention. To explain the redshift of distant galaxies, the temporalistic model does not therefore need the effect of the various expanding Universe (FLRW) cosmological models.

The interpretation by the Big Bang model of the redshift of distant galaxies as being a cosmological effect caused by the expansion of space, is also refuted by the temporalistic model. The cosmological effect $z = vr / c$ is interpreted in the temporalistic model by $z = t / To$, where z is the redshift, vr the apparent radial velocity, c the speed of light, t the travel time of the photon (or distance / c), and To the temporalistic constant.

Whereas, in the Big Bang model, expansion only begins beyond the local system of galaxies, in the temporalistic model, redshift (or recession effect) takes place as soon as the photon is emitted.

In calculating the redshift or 'recession effect', we have not taken into account relativistic correction. At high, or more specifically, relativistic speeds, in other words close to the speed of light, redshift and the 'recession effect' are different, as can be observed in the spectra of distant quasars. The shift in wavelength can be in the region of several times its original value, and the 'recession effect' several times c.

The relativistic correction of redshifts and of the recession speeds of distant galaxies applies in the expanding Universe. This is due to the upper limit on the speed of light, a premise accepted in the expanding Universe model as well as in the temporalistic model, and the resulting slowing down of clocks. However, relativistic correction cannot play a role in the temporalistic Universe because it concerns light sources moving at relativistic speeds. In the temporalistic model, it is the radiation that varies, while the galaxies are stationary. Here, the 'recession effect' is an apparent effect and does not correspond to a cosmological effect at relativistic speeds. The relativistic redshift, at large distances or over large periods of time, nonetheless remains an experimental fact, and one which cannot be explained in the temporalistic model by a relativistic effect because in this model the light sources are stationary. So how are we to interpret it, then, in the temporalistic model?

In the expansion model, the redshift z at non-relativistic speeds due to the radial cosmological effect is given by the equation $z = vr/c$. c is a speed that in a vacuum cannot be exceeded by any other physical speed. It is a limiting constant. In the temporalistic model, the constant To is, similarly, a limiting constant for time periods. The redshift over short time periods is given by the equation $z = t / To$.

Temporalistic redshifts, over temporalistic time periods, are similar to relativistic redshifts at relativistic speeds. The essential difference between the relativistic redshift and the temporalistic redshift stems from the origin of the redshift: On the one hand, a factor that is external to the radiation, the expansion of space, and on the other, the quantum temporalistic effect that is internal to the radiation. The new explanation of the redshift z of distant galaxies put forward by the temporalistic model naturally has major cosmological implications.

At a distance of 14.43 billion light years, after temporalistic correction, wavelengths and the 'recession effect' become infinite, which implies a cutoff point in observable space. Beyond this limit, the Universe, which physically continues in space, is no longer accessible to us. This is the temporalistic horizon. In the expanding Universe model, we end up with a cosmic horizon of the same order of magnitude, but this horizon is one of space, whereas the temporalistic horizon is one of time. The only limits of the Universe for the observer are those imposed by the redshift of electromagnetic radiation induced by the temporalistic constant, i.e. 4.55456×10^{17} s in time, and around 13.65×10^{25} m in space. However, it would be hazardous to assert that the limits of the observable Universe are the same thing as the limits of the Universe.

To sum up, the redshifts of distant galaxies are quantum phenomena that are the result of the existence of the temporalistic constant To, and not the result of macroscopic cosmological phenomena leading to a model of expansion and to the Big Bang.

Without making any other assumptions, the temporalistic model naturally leads to the proposition of gravitation with a finite range (< http://site.voila.fr/nobigbang > chapter 9).

See calculations: CHAPTER XV

page 181

Chapter VIII

CONSEQUENCES AND WEAKNESSES OF THE SPATIAL INTERPRETATION OF REDSHIFTS

THE THREE PILLARS OF THE STANDARD BIG BANG MODEL

The standard Big Bang model is based on a certain number of pieces of evidence, or so-called evidence, of which the most important, according to the scientists who support the theory, are called the 'Three pillars' of the standard Big Bang model. These three pieces of evidence are redshifts (of distant galaxies), the cosmic microwave background, and primordial nucleosynthesis.

We shall first analyze these 'three pillars'. We shall see that, contrary to the peremptory assertions of the Big Bang model's supporters, 1) this 'evidence' is nothing of the sort, 2) they are in fact mere hypotheses, and 3) these hypotheses are not only weak but in addition are backed up by highly questionable arguments.

We shall see that the first and most important piece of 'evidence', redshifts, far from providing a foundation or a 'pillar' for the standard Big Bang model, is actually completely detrimental to it.

The redshift argument which, historically, was the starting point for the Big Bang model, has unfortunately led cosmologists up the wrong track, and consequently, as we shall see, has given rise to considerable, unresolved problems which concern every concept in the standard Big Bang model.

1 Redshifts

According to the standard Big Bang model:

In the 1920s, observations of the spectra of distant galaxies by Vesto Slipher and Hubble showed that the spectra of these galaxies were shifted in comparison with the same spectra observed in the laboratory. In 1929, Hubble showed that the redshifts of galaxies became greater the further away they were. He concluded that all the galaxies were moving away from the Earth, and that this was a Doppler-Fizeau type effect. The redshift z did not exceed 0.007%. Together with Milton Humason, Hubble formulated Hubble's law: $v = Ho \times d$, where v is the recession speed, Ho the Hubble constant, and d the distance of the galaxy.

See calculations –Chapter XV – page 181

The recession speed of galaxies estimated by Hubble at the time was in the region of 500 km/sec/Mpc (Ho) and the 'age' of the Universe (to) around 2 billion years.

In 1963, a radio source was discovered, quasar 3C 273. Its optical spectrum was redshifted by 0.158%. The value of Ho (the constant of proportionality of the apparent recession speed of galaxies) is currently estimated to be around 68 km/sec/Mpc.

Numerous quasars have since been discovered, and at present a redshift z of around 10 has been attained. The redshift of the cosmic microwave background is currently estimated to be $z = 1100$.

Since the very first estimations of the Hubble constant in 1929 (Ho = 500 km/sec/Mpc and to = 2 billion years), the standard Big Bang model has, over the decades and after many adjustments and very numerous observations, arrived at roughly the values established <u>theoretically in 1962</u> by the temporalistic model, i.e. Ho = 67.71 km/sec/Mpc and To = 4.5546 × 10^{17} s (approximately 14,43 billion years).

Interpretations of the redshifts of distant galaxies can mainly be attributed to:

1) the Doppler-Fizeau effect (the source of a wave is moving in relation to the observer);

2) gravitational redshift (variation in the energy of light when it is subjected to a gravitational field, or Einstein effect);

3) cosmological redshift, interpreted as being an effect caused by the expansion of the Universe or by the curvature of the Universe.

Many other explanations and interpretations have been put forward, such as variation in the speed of light, the concept of 'tired light', etc. The only interpretation accepted by prevailing cosmological models and the standard Big Bang model is that of the expansion of the Universe.

Critiques:

As we pointed out a little earlier, the values established theoretically in 1962 by the temporalistic model for the Hubble constant are Ho = 67.71 km/sec/Mpc, and To = 4.5546 × 10^{17} s (around 14.43 billion years). These values were confirmed on 25 July 2011 by Florian Beutler, ICRAR – UWA (University of Western Australia), according to whom "the new measurement of the Hubble constant is 67.0 ± 3.2 km s^{-1} Mpc^{-1}", in other words within 1% of the temporalistic value of Ho, 67.71 km s^{-1} Mpc^{-1}.

Contrary to the customary assertions of the supporters of the standard Big Bang model, the redshifts of distant galaxies are not evidence for the Big Bang model. This is merely one interpretation of cosmological observations. These observations are therefore simply hypotheses interpreted in a way that is favorable to another hypothesis, the standard Big Bang model. Many other interpretations and hypotheses have been proposed, but none have been accepted. This does not imply that the Big Bang hypothesis is true: it is only accepted 'by default'.

The temporalistic model proposes a new alternative to the Big Bang hypothesis.

The interpretation (hypothesis) of the redshifts of distant galaxies by the standard Big Bang model leads to numerous (factual and theoretical) anomalies and problems for this model. Let us recall the main ones. The concepts of space and time, as well as the concept of the expansion of the Universe, are concepts where geometry and physics are combined and confused with each other. For instance, the expansion of the Universe is explained as being an expansion of space, which may have mathematical validity if the concept is consistent, but in no way implies its physical validity. On the other hand, the expansion of space considered as a physical concept, in other words as the expansion of the container of phenomena and events, namely the vacuum, is a completely contradictory concept. Space and spacetime are geometrical concepts. To attribute them to the empty physical Universe, i.e. to nothingness, is to endow it with properties that it cannot possess, by definition. Our analysis suggests that this is a purely anthropic concept.

The interpretation of redshifts by the standard Big Bang model as being caused by the expansion of space has naturally and inevitably brought with it many other theoretical and practical problems that the temporalistic model does not have: the origin of the Universe, the primordial singularity, the original infinite values of temperature, density, space, etc ... magnetic monopoles, the cause of the primordial explosion, the many problems attached to the homogeneity and flatness of the Universe, primordial nucleosynthesis, critical density, the age of the Universe and its various structures (stars, clouds of matter, galaxies, clusters and superclusters of galaxies, filamentous structures, large-scale structures and walls, great voids, etc), dark matter, dark energy together with the many hypotheses that attempt to explain it and the problems attached to them (cosmological constant, vacuum energy, quintessence, etc), highly speculative inflationary hypotheses which, without any observational justification, constantly infringe current physical laws (exponential speed of events − inflation − sudden, arbitrary beginning and end of inflation − unjustifiable and unjustified − spatial and temporal phenomena − dark energy, etc). Many other difficulties, resulting directly from the erroneous premises of the interpretation of redshifts as being caused by the expansion of space, form an impressive body of evidence pointing to the incorrect nature of the current dogma of the Big Bang model.

We will now undertake the critical analysis of a large number of concepts in the standard model of cosmology and of the weaknesses that we have just mentioned, which are the inevitable consequences of the point of departure (the premises) of a model based on erroneous foundations, the spatial interpretation of the redshifts of distant galaxies, which we earlier called the 'original sin' of the standard Big Bang model.

As we saw earlier, the dominant concept in the standard model of cosmology, the expansion of space, which results from an incorrect interpretation of the redshifts of distant galaxies, is an anthropic concept, which, in an inconsistent manner, confuses the empty spatial framework, the container, with the content (the various characteristics of the Universe, atoms, clouds of dust, large and small structures, planets, stars, galaxies, clusters and superclusters of galaxies, great voids, etc). This interpretation, derived from the equations of general relativity, can claim to be valid mathematically, based on its consistency. However, in no way does this mean that it is valid physically. General relativity modifies geometry and distorts spacetime in Minkowski's special relativity by providing it with curvature. In general relativity, spacetime is a manifold whose curvature is identified with gravitation. General relativity appears to give space, or rather spacetime, almost material properties. How can a <u>physical</u> space be curved by mass-energy? If it is a vacuum, then this concept is contradictory. Nothingness cannot be curved. If it is not a vacuum, then the reality of this <u>physical</u> space is unclear. It is neither a vacuum nor is it the content of the Universe (mass-energy). As can be seen, there is a constant shift between the geometrical and physical concepts of space, between the container and the content of the Universe. The curvature of spacetime brought about by Einsteinian gravitation should be interpreted as being a curvature of geodesics, i.e. of the paths of test particles, matter, stars, galaxies, etc moving through the Universe, rather than as a curvature of the container (space or vacuum). This <u>physical</u> interpretation of Einsteinian gravitation alone is consistent, non-contradictory and therefore ananthropic. In Chapter XII we will see how this interpretation fits in with temporalistic gravitation.

It is well worth recalling that, although the expansion of the Universe is presented as resulting from the equations of General relativity, Albert Einstein did not go along with this, and even attempted to put forward an alternative, the concept of 'tired light'. Many years later, he was forced to give up this concept.

It is also important to realize that the interpretation of redshifts as being caused by the expansion of space played a major role in the elaboration of the Big Bang model. This interpretation is what we called the '<u>original sin</u>', on the basis of which the erroneous premises of the Big Bang model have led the vast majority of researchers up the wrong track, as we shall see in this analysis.

The various phenomena or concepts that belong to the standard model of cosmology, the Big Bang model, are presented by the researchers who support this model as evidence of its validity. We shall see that the supporters of the Big Bang model usually put forward this so-called '<u>evidence</u>' for the model, while totally eluding its difficulties or any contradictory evidence. We will see that in the majority of cases the interpretation of facts replaces their neutral or ananthropic observation: the interpretation of the expansion of the Universe replaces the observation of redshifts, the interpretation of fossil radiation and of fluctuations in the density of the primordial Universe replaces the observation of the cosmic microwave background and of fluctuations in the temperature of the cosmic microwave background, etc.

The two other concepts in the standard Big Bang model, presented as pillars of the theory are the cosmic microwave background and primordial nucleosynthesis. Our critical analysis of these two concepts will make it clear that, just as for the concept of redshifts, these two concepts, that are put forward as evidence of the Big Bang, rely on an interpretation of the facts rather than on neutral observation of them, without forgetting other, more specific problems.

2) The Cosmic Microwave Background (CMB)

According to the standard Big Bang model:

The fluctuations in the cosmic microwave background were observed by the COBE (Cosmic Background Explorer) satellite in 1992, then by the Archeops, Boomerang, Maxima and TopHat balloons, followed by the WMAP 5 spacecraft (Wilkinson Microwave Anisotropy Probe 2003 – 2008). The precision of the Planck spacecraft, which was launched in May 2009, is 50 times greater than that of WMAP 5.

The existence of the cosmic background radiation was predicted in 1940 by Ralph Alpher, Robert Herman and George Gamow as a consequence of the Big Bang model. They predicted it again in 1949.

In 1964, Arno Penzias and Robert Wilson discovered the cosmic microwave background, i.e. the microwave radiation which uniformly fills the Universe. The cosmic microwave background is black body radiation at a mean temperature of 2.725 kelvins.

This research has produced the following findings:

1) the cosmic microwave background is fossil radiation dating back to 380 000 years after the Big Bang

2) dipole anisotropy of the radiation, resulting from the Doppler effect caused by the motion of the Earth

3) minimal fluctuations in the radiation, with fluctuations in the region of 10^{-5} K around the mean temperature of 2.725 K

4) the determination of the radiation's origin by studying its polarizations, either inflationary or topological (with the effect of gravitational waves and the possibility of the existence of cosmic strings)

5) a model for the formation of the structures of galaxies in the Universe

6) the origin of the formation of the first structures of galaxies

7) the probable composition of the Universe: 4% baryonic matter, 24% dark matter, 72% dark energy accelerating the expansion (cosmological constant? quintessence?)

8) the Universe is nearly spatially flat.

Critiques:

1) Contrary to the historical version put around by the supporters of the Big Bang, the existence of the cosmic microwave background does not result from the Big Bang model alone. It was predicted, without using the Big Bang model, and often well before Gamow, by Guillaume (1896), Eddington (1926), Regener (1933), Nernst (1933), McKellar and Herzberg (1941), Finlay-Freundlich (1953) and Max Born (1953). These authors predicted temperatures ranging from 1.9 K to 6 K (André Koch Torre Assis and Marcos Cesar Danhoni Neves - 1995). In addition, Gamow's prediction in 1953 of a cosmic background radiation at a temperature of 7 kelvins was based on a fallacious mathematical argument (Weinberg 1980).

2) The cosmic microwave background is not evidence for the existence of the Big Bang. Once again, it is merely an interpretation, correlated with a hypothetical model, the Big Bang model, of a factual phenomenon. Many other interpretations are possible, as we pointed out in the preceding paragraph. The interpretation of the cosmic microwave background is not, therefore, evidence for the Big Bang model but a new hypothesis which has

no more validity than the hypothesis of the expansion of space used to explain redshifts.

3) According to the Big Bang model, the cosmic microwave background is interpreted as being fossil radiation dating back 13.7 billion years. The variations in the temperature of the cosmic microwave background are in the region of a few tens of microkelvins, i.e. 1 in 100 000. These fluctuations in the temperature of the cosmic microwave background are interpreted as being fluctuations in the density of the primordial Universe, and as being the seeds from which the galaxies and the other large-scale structures of matter in the Universe formed. No evidence validates this interpretation, which, once again, can be seen to be a simple hypothesis rather than evidence.

4) The use of the cosmic microwave background as a validation of the hypothesis of the existence of the Big Bang poses a number of problems:

a) the quasi-uniformity of the cosmic microwave background poses the problem of the horizon: why are regions of the Universe that are too far apart to have ever been in contact via signals traveling at the speed of light at almost exactly the same temperature? The standard Big Bang model cannot and does not answer this question. To try and get round this problem, this later gave rise to the emergence of a new hypothesis (see Chapter V, above, and Chapter VIII – Inflationary theories).

b) the small size of the fluctuations in the cosmic microwave background is not enough to justify quantitatively the formation of the galaxies and the large-scale structures of the Universe.

c) no justification is offered for the fact that the flatness of the Universe is almost equal to its critical density

d) why is the Universe homogeneous and isotropic?

5) The various problems raised by the interpretation of the cosmic microwave background as being so-called evidence for the phenomenon of the Big Bang have led to new *ad hoc* hypotheses, the inflationary models, which have no experimental or factual foundation and provide responses that are highly and exclusively speculative, as we shall see in Chapter VIII (Inflationary theories).

6) As time goes by, other specific problems have been added on to these general problems. For instance, in January 2009, the Arcade Mission led by NASA's Alan Kogut discovered that the cosmic radio background left by

the 'reionization' of the cosmos at the end of the 'dark ages' (a hypothetical concept that is specific to the Big Bang model) is six times stronger than predicted. To date, no satisfactory explanation has been found for this discrepancy between the evidence and this hypothetical concept in the Big bang model.

7) Another recent problem in the Big Bang model for which there is no explanation: an international team of astronomers which has been studying a gamma-ray burst using the European Southern Observatory's VLT (Monthly Notices of the Royal Astronomical Society, 2 November 2011), has observed that the absorption lines of two very young galaxies (12 billion years old), through which the gamma-ray burst travelled, show that they are very rich in heavy chemical elements, which clearly contradicts the predictions of the Big Bang model.

To sum up, the explanation of the cosmic microwave background and the hypothetical inflationary theories connected to it are more similar to Ptolemaian epicycle-type reasoning than to a rigorous scientific model that respects current physical laws and principles (law of conservation of mass-energy, 'validated' or 'falsifiable' hypotheses, etc), i.e. to ananthropic concepts and propositions.

3 Primordial nucleosynthesis

According to the standard Big Bang model:

One of the major arguments for the Big Bang model is the synthesis of light elements a few minutes after the Big Bang. This is the standard Big Bang Nucleosynthesis model. This primordial nucleosynthesis, characterized by the primordial abundances of light elements, depends on the initial conditions of a single free parameter, the baryon/photon ratio Eta. This ratio is currently estimated to be between 4.5 and 4.9×10^{-10} (Trento 1997).

In the first three minutes following the Big Bang, the nuclei of light elements were created from baryons: ^{2}D, ^{3}He, ^{4}He and ^{7}Li (Weinberg 1980). The current abundances of these elements, compared to hydrogen are: ^{2}D = 4.9499624 × 10^{-5}, ^{3}He = 1.3265581 × 10^{-5}, ^{4}He = 0.24387701, ^{7}Li = 1.8648816 × 10^{-10} (Craig Hogan - Luis Mendoza 1998). The heavier elements were subsequently created inside stars.

The agreement between the predictions of abundances of light nuclei using the basic hypotheses of the Big Bang and current abundances of these nuclei is one of the model's strong points. However, it should be pointed out that there are a large number of versions of non-standard Big Bang scenarios. Thousands of articles have been devoted to them. They are based on initial conditions for the Big Bang that are different from the standard model (mainly the baryon/photon ratio, but also other hypotheses such as inhomogeneities, non-standard properties of neutrinos, etc). Nonetheless, all these models are based on the Big Bang model, but with different initial conditions.

Critiques:

1) The very name of this concept, both in French and English ('explosion primordiale' and Big Bang Nucleosynthesis), constitutes a prior assumption that starts out from the current abundances of the chemical elements, which are facts, and infers from these a hypothetical primordial nucleosynthesis. Therefore, in no way is this evidence, but rather an interpretation, which needs to be validated, of actual established facts.

2) The origin of the creation of lithium a few minutes after the Big Bang is subject to debate. An unexpected origin for lithium has been discovered in red giants in around twelve globular star clusters. It may result from the decay of the unstable radioactive isotope beryllium-7. Lithium is also found in other very massive red giants at a late stage in their evolution (Catherine Pilachowski 2001). Moreover, the amount of lithium produced before the stars formed and the amount destroyed in stars is unknown.

3) There is a discrepancy between the values predicted by Big Bang nucleosynthesis for deuterium, ^{2}D, and observations by researchers (Trento 1997). The figures for ^{4}He are 0.246 ± 0.0014 for Burles and Tytler and 0.234 ± 0.002 for Olive, Steigman and Skillman (OSS 1999). For the

baryon/photon ratio, the discrepancies range from $5.1 \pm 0.5 \times 10^{-10}$ (Burles and Tytler) and $2.1 \pm 0.6 \times 10^{-10}$ (OSS).

4) According to the standard cosmological model, the value of baryonic density, a few seconds after the Big Bang, was between 3×10^{-10} and 5×10^{-10}.

According to the map of fluctuations observed by the Boomerang collaboration (2000), the baryonic density in the cosmic background radiation dating from 380 000 years after the Big Bang was 7.4×10^{-10}. This discrepancy calls into question the standard Big Bang model for nucleosynthesis.

All the hydrogen and part of the helium and lithium found in the Universe are supposed to have been formed in the first hundred seconds after the Big Bang. Astrophysicists pay close attention to primordial nucleosynthesis. This is because the slightest result that refutes its predictions threatens the models of the Big Bang.

A comparison of the results from the latest theoretical calculations for nucleosynthesis and from WMAP 5 data show that the values inferred from the cosmic microwave background and astrophysical observations <u>agree for deuterium, are no more than reasonable for helium-4, but are in complete disagreement for lithium-7.</u>

- :-
-

The three series of phenomena described above, 1) redshifts, 2) the cosmic microwave background, and 3) primordial nucleosynthesis, are generally considered to be the <u>three key pillars</u> that support the standard model of cosmology. We have just seen that, according to our analysis, these three pillars, far from being set in stone, rest on quicksand.

The redshifts of distant galaxies are interpreted as being caused by the expansion of the Universe. Such concepts are not observational data. They result from an interpretation, i.e. a hypothesis based on confusion between the <u>geometrical</u> and <u>physical</u> concepts of spacetime. The temporalistic model proposes an alternative, validated interpretation of redshifts.

The interpretation of the cosmic microwave background as being evidence for the Big Bang model, far from supporting this theory, piles up problems for the Big Bang concept, as we saw earlier: problems relating to the horizon, to the flatness of the Universe, to critical density, to the homogeneity and isotropy of the Universe, etc.

The numerous difficulties for primordial nucleosynthesis that we mentioned earlier, including the major discrepancy concerning the abundance of lithium-7, are crippling for the validity of the primordial nucleosynthesis concept, a hypothesis that results directly from the Big Bang paradigm.

It should be noted that, unfortunately, when the Big Bang model is expounded, the many problems and difficulties that beset the theory are usually neglected: the so-called 'evidence' of redshifts, which in fact is merely an interpretation, the prediction of the cosmic microwave background by many researchers, well before Gamow and his prediction based on a fallacious mathematical argument (Weinberg 1980), the many discrepancies in the primordial nucleosynthesis concept that are disregarded, etc.

The redshifts of distant galaxies, interpreted in the standard Big Bang model as being caused by the expansion of the Universe, are the premises from which all the concepts of the Big Bang result. If the temporalistic alternative is right, we shall see, in the following analyses, that the concepts relating to the Big Bang model are inevitably affected by the problems of this model, and that these new problems then lead to new arbitrary, *ad hoc* hypotheses that have no experimental or factual foundation or validation, such as speculative inflationary theories, which completely disregard the current laws of physics.

- :-

Inflationary theories

According to the standard Big Bang model:

As a consequence of the difficulties encountered by the Big Bang model, many fresh problems have arisen: the horizon problem: why are regions of the Universe that are too far apart to have ever been in contact via signals traveling at the speed of light at almost exactly the same temperature? ; the flatness of the Universe being almost equal to its critical density has no justification; why is the Universe homogeneous and isotropic? This is what led to the creation of *ad hoc* and entirely speculative hypotheses, the inflationary theories, which have come in various versions: the theory of inflation elaborated by Alexei Starobinsky was developed by Allan H. Guth and Paul Steinhardt (1984 – 1998), Andy Albrecht, Andrei Linde (1994 – 2001).

According to theories of inflation, the visible Universe originated in a very small, very hot (10^{32} kelvins) region of the homogeneous Universe which inflated 10^{-35} seconds after the Big Bang. This inflationary phase lasted 10^{-32} seconds during which the Universe expanded by a factor of around 10^{50}, after which the Big Bang continued to evolve. The explosion is supposed to have been a consequence of the energy density of the vacuum, which gave rise to repulsive gravity caused by the existence of the cosmological constant, Λ (Einstein initially added this constant to his equations, before later rejecting it).

The theory of inflation has the virtue of solving a certain number of problems raised by the Big Bang model:

1) The extraordinarily rapid inflation of the Universe, at speeds well over the speed of light, originating in a tiny, homogeneous region of the Universe, solves the horizon problem.

2) A flat Universe with a density close to the critical density also results from the inflationary model.

3) The problem of magnetic monopoles: for the creation of nuclei in the primordial Universe, the Big Bang model requires the use of the Grand Unification Theory (GUT) and the production of massive particles, called magnetic monopoles. Many of these magnetic monopoles should still be around today. So where are they? Once more, the answer to this question is provided by inflationary theories. The lack of any magnetic monopoles today is explained by their rapid dispersal during the inflationary phase.

4) The inflationary model predicts small fluctuations in the cosmic microwave background (in the region of 10^{-5}), leading to the formation of galaxies.

Critiques:

1) The theory of inflation is an extension of the Big Bang model, but it is independent of it.

2) Inflationary models, created in order to solve the problems of the Big Bang model, are not based on any experimental or factual evidence. The considerable extrapolation of the laws of physics set out in these models has no theoretical justification, apart from providing an arbitrary response to the difficulties of the Big Bang model. In the final analysis, it is nothing more than an *ad hoc* hypothesis. "The assertions of the inflationary model, which are poorly demonstrated, can lead to genuine scepticism in the eyes of rigorous observers" (Peebles 2001). Incidentally, piling up hypotheses without any observational basis, can lead to highly speculative and particularly questionable versions of inflation: chaotic inflation, self-reproducing universes, multiple universes, parallel universes, bubble universes with eternal inflation, creation of universes in a laboratory, creation of universes by a physicist-hacker, and other wild ideas light years away from the required scientific rigor!

3) The very small fluctuations in the cosmic microwave background do not satisfactorily explain the formation of the large-scale structures of the Universe (galaxies, clusters and superclusters, great walls, voids, etc).

4) The existence of the cosmological constant, proposed and then rejected by Einstein, which is required by inflationary models, remains a pure hypothesis at present, just like all the assertions of inflationary models. Later on we shall see that the concept of the cosmological constant leads to insurmountable difficulties with regard to physical reality.

5) The cause of inflation, which began when three out of the four fundamental interactions had dissociated, remains unknown.

6) The beginning and then the end of inflation are only justified by means of further hypotheses.

7) A major argument that leads to the rejection of inflationary theories is their ability to fit all possible initial conditions. According to Peebles, an eminent supporter of the Big Bang: "It (inflation) is a theory that can be

adjusted in order to produce the structures that we see starting from all the possible initial conditions. In this sense, it is not truly a theory, but an 'off the peg' story, since it is suitable in all cases. All that is necessary is to change a few parameters." "In any case, we don't have a better one (inflation mechanism)." (Dossier trimestriel N° 35- Mai 2009 - La Recherche – page 8). This is a resigned acceptance 'by default'.

8) To get round the difficulties of the Big Bang model, cosmologists have transferred them to another, even more hypothetical concept, inflation, which is a genuinely Ptolemaian approach. Incontestably, this is to jump out of the frying pan into the fire.

The origin of the Big Bang

According to the standard Big Bang model:

The Big Bang universe expanded from a spatio-temporal singularity which must have been infinitely dense. As we have just seen, according to theories of inflation, the visible Universe originated in a very small, very hot (10^{32} kelvins) region of the homogeneous Universe which inflated 10^{-35} seconds after the Big Bang. This inflationary phase lasted 10^{-32} seconds during which the Universe expanded by a factor of around 10^{50}, after which the Big Bang continued to evolve.

The most distant observations of the Universe accessible to telescopes are located 380 000 years after the Big Bang, in other words at the time the cosmic microwave background radiation was emitted. Moreover, it is not possible to go further back than the Planck time (10^{-43} seconds after the Big Bang), since the equations of both general relativity and of quantum field theory can no longer be used due to the appearance of many infinite terms. The latest data provided by WMAP 5 (Table 7 – Cosmological Parameter Summary – 2008) gives a value for Ho = 71.9 (+2.6 – 2.7) km/s/Mpc and to = 13.69 (± 0.13) billion years. The Big Bang brings about the appearance of space and time, or of spacetime, as well as of mass-energy. Since time was created at the same time as the Big Bang, it is impossible to go back any further, in other words beyond 13.7 billion years.

Critiques:

1) According to the Big Bang model, the Universe was born, in the 'primordial explosion', from a spacetime singularity which had an 'infinite' density and temperature. What was the cause of this explosion? No answer to this question is provided by the current laws of physics. Or else, this difficulty is side-stepped by denying the 'primordial explosion', without any clear or valid justification. Where did space, time, matter and energy come from? They were created *ex nihilo*, equally without any experimental or factual validation. And yet, until now, no *ex nihilo* creation of matter or energy has ever been observed, whether in physical or biological phenomena.

2) To assert that space and time appeared with the Big Bang is a circular argument which obviously cleverly eliminates, without any validation, the problem of the existence of time before the Big Bang.

3) It is also equally possible to assert, without any more justification, that at the singularity of the Big Bang, the notion of space disappears but not that of time (Gabriele Veneziano's pre-Big Bang). As usual, this new hypothesis is neither 'verifiable' or 'falsifiable'.

4) Many other hypotheses have been put forward: the ekpyrotic model, which proposes a multidimensional brane universe where inflation is replaced by the cyclic collision between two universes; the strictly speculative model of eternal inflation of bubble-Universes, etc.

5) All these models without exception are speculative, without there being any possibility of validating them. However, this does not appear to bother their authors, who loudly claim the right to speculate without any restrictive tests of validity (Andrei Linde).

6) When all is said and done, the Big Bang model is a strictly anthropic concept. It violates several of the ananthropic criteria that we set out in Chapter III: it is irrational and speculative at the expense of critical thinking; it infringes current physical laws without providing any experimental, observational validation; it uses contradictory concepts such as a 'quantum vacuum filled with quantum fluctuations', etc.

<u>The acceleration of expansion – Dark energy</u>

According to the standard Big Bang model:

1) After the inflationary phase at 10^{-35} s, the size of the Universe was multiplied by 10^{50}, and then for the following 8 billion years the expansion of the Universe was slowed down by gravitation. In 1998, the observation of the redshifts of type Ia supernovae led researchers to recognize an acceleration of expansion, roughly 5 billion years ago, assuming a standard mechanism of formation for type Ia galaxies. The acceleration of expansion is attributed to a mysterious dark energy, and is based on the hypothesis of a homogeneous Universe, with a homogeneous and isotropic distribution of matter, statistically speaking (Copernican principle).

2) Evidence for the acceleration of the expansion of the Universe: Type Ia supernovae, galaxy cluster counts, the effect of gravitational lenses, and evidence for the existence of dark energy: type Ia supernovae, the cosmic microwave background (fluctuations) directly correlated to the geometry of the Universe (flat according to Boomerang), followed by WMAP 5 and acoustic waves.

3) Models for the nature of dark energy: a) the cosmological constant, Λ, likened to vacuum energy; according to the predictions of quantum field theory, with quantum vacuum fluctuations. In quantum field theory, the vacuum is not nothingness, but rather the fundamental minimum energy state of quantum field systems; b) quintessence; c) modified general relativity, MOND; d) axions, the transformation of some photons into axions, which go undetected by telescopes, thus leading to an underestimation of the luminosity of galaxies, which is interpreted as being an acceleration of expansion.

4) Dark energy introduced into the standard Big Bang model in the form of a cosmological constant.

Critiques:

1) The formation of type Ia supernovae appears to be varied than previously believed. "Type Ia supernovae are used to evaluate the expansion of the Universe assuming a standard mechanism of formation." However, there is nothing standard about the formation of type Ia supernovae (supernova SN2006gz), and this distorts cosmologists' measurements (Stéphane Fay – Astrophysical Journal Letters, vol. 669 pp.L17-L19.2007).

Thomas Buchert (University of Lyon) and David Wiltshire (University of Canterbury) "criticize the relationship between the redshift of astronomical objects (especially supernovae) and their distances, which is itself based on the hypothesis of a homogeneous Universe. A mechanism explaining how inhomogeneities would modify the propagation of light on very large scales has therefore been proposed."

"The acceleration of expansion might only be the consequence of a wrong symmetry hypothesis. Several teams have shown that certain models without accelerated expansion could reproduce the observations of supernovae if it is assumed that we live in a reduced density region of the Universe, a sort of bubble whose density is lower", Jean-Philippe Uzan (Dossiers La Recherche - May 2009 – p 91).

2) The acceleration of expansion leads to the hypothesis of the existence of dark energy. Different models propose an explanation for this: a) the cosmological constant introduced by Einstein and then rejected by him (according to Einstein, it was the greatest blunder in his life), and considered similar to vacuum energy. Unfortunately, the predictions of quantum field theory give a totally unacceptable value, 60 to 120 times greater than the value inferred from cosmological observations. This value, inferred from quantum field theory, and incompatible with the properties of the Universe, constitutes a major conceptual problem that is still unresolved; b) quintessence (very popular a few years ago, but abandoned since because of the many problems it raises; c) general relativity with its 'scalar tensors'; no observation has validated this concept, which remains a pure hypothesis; d) axions, particles that result from the transformation of a certain proportion of photons: this model has today been abandoned. Many of these models (such as quintessence) have 'free functions' that can be fitted to those of the cosmological constant, thus making them impossible to refute and therefore not 'falsifiable'. None of the models proposed has therefore been validated. In desperation, some have had no hesitation in proposing an 'anthropic model'!

3) The concept of quantum vacuum fluctuations in quantum field theory is an oxymoron. This concept, whose operational value cannot be denied, is theoretically inconsistent. A vacuum (nothingness) cannot, by definition, have properties or contain anything at all, even if it is virtual. If not, it is a non 'falsifiable', i.e. 'anthropic' concept. As we pointed out in the previous paragraph, no model for the nature of dark energy has been validated.

4) Dark energy, introduced into the standard Big Bang model in the form of a cosmological constant, Λ, suffers from the same difficulties as this latter concept.

5) Only inhomogeneous and non-isotropic models of the Universe, with their questioning of the cosmological principle, evade criticism. They lead to a rejection of the acceleration of expansion and its consequence, the existence of dark energy.

6) According to the latest research by Arman Shafieloo and colleagues (14 April 2009), concerning nearby supernovae (less than a billion light years away), the acceleration of expansion has diminished over the last 2.5 billion years, to the point of reversing recently. This implies a similar fall in the density of dark energy, which would mean the exclusion of the cosmological 'constant' Λ (http://arxiv.org/abs/0903.5141).

To sum up: the concepts of expansion, acceleration of expansion and dark energy, together with all the problems they bring about, are the direct consequence of the spatial interpretation of the redshifts of distant galaxies.

By refuting any idea of expansion of space, the temporalistic interpretation of redshifts naturally sidesteps all these problems and the puzzle of the existence of dark energy.

Various problems

The various theories of the Big Bang come up against a certain number of theoretical and factual problems:

1) the horizon problem
2) the problem of flatness and critical density
3) the singularity problem

4) the problem of the homogeneous and isotropic Universe

1) The horizon problem

According to the standard Big Bang model:

Observations of the cosmic microwave background show that, on large scales, the Universe is homogeneous and isotropic (with a precision in the region of 10^{-5}). The Friedmann equations show that, at a given time, a Universe that is homogeneous and isotropic remains so. Before inflation, the regions of the Universe, which were still very close together, had 'plenty of time' to exchange their properties (such as temperature). With inflation, these neighboring regions moved away from each other. Expansion was a local phenomenon which took place homogeneously at every point in the primordial Universe. This model is a representation in time and not in space. It is reasonable to suppose that, shortly after the Big Bang, all the matter observed was located in a small region, so that it can be assumed that it was homogeneous and isotropic, and that the Universe then underwent a period of exponential expansion (inflation) which moved the different regions in this area away from each other very rapidly. However, it is very difficult to explain why, right from the start, the Universe ended up being homogeneous and isotropic.

The solution is inflation, which when it replaces normal expansion, enables an exponential expansion of the Universe without violating the speed limit of special relativity. This solution is possible, according to the Friedmann equations, by assuming that a form of matter that has negative pressure exists in the Universe.

Critiques:

1) In order to solve this new problem, the standard Big Bang model of cosmology requires a new hypothesis, the highly speculative hypothesis of inflation, a non 'falsifiable' hypothesis whose many difficulties were

analyzed above and which, far from solving the horizon problem, simply piles up fresh problems.

2) The solution to the horizon problem which consists in postulating the existence in the Universe of a form of matter that has negative pressure, for which there is no experimental or observational validation, merely constitutes another *ad hoc* hypothesis without any theoretical justification.

3) The supporters of the Big Bang model admit: "However, it is very difficult to explain why, right from the start, the Universe ended up being homogeneous and isotropic."

4) The hypothesis of the existence in the Universe of a form of matter that has negative pressure brings us back to the cosmological constant Λ, considered similar to vacuum energy, which is known to result from the predictions of quantum field theory, leading to a totally unacceptable value, 60 to 120 times greater than the value inferred from cosmological observations.

2) The problem of flatness and critical density

According to the standard Big Bang model:

Observations show that the Universe is almost completely flat, with an energy density of the same order of magnitude as the critical density corresponding to a universe with zero spatial curvature. Why should this be? The energy density of the Universe could take any value. The Big Bang model provides no justification for this flatness.

The solution is the same paradigm that provides a solution to the horizon problem: inflation. If inflation increases the size of the Universe by a factor of 10^{50}, its curvature is reduced by an identical factor. Its current value is therefore very close to zero and its energy density very close to the critical density.

Critiques:

1) Inflationary theories, which are highly speculative, are supposed to solve the flatness problem even though they are themselves a source of serious difficulties. The solution proposed for the flatness problem, which is identical to the solution to the horizon problem, namely inflation, therefore suffers from the same difficulties, in other words it is a highly speculative hypothesis, with the *ad hoc* concept of matter with negative pressure, lacking any observational backing, and with a totally unacceptable value, 60 to 120 times greater than the value inferred from cosmological observations

2) The supporters of the Big Bang model are forced to admit yet again: "The Big Bang model provides no justification for this flatness."

3) The opinion of a renowned theoretical cosmologist and supporter of the Big Bang, James Peebles, about inflationary theory, which we quoted earlier, and which also applies to this problem, is very critical and instructive (see "Inflationary theories", Peebles p 87).

3) <u>The singularity problem</u>

<u>According to the standard Big Bang model:</u>

In the 1960s, Stephen Hawking and Roger Penrose's 'singularity theorems' showed the inevitable presence of a cosmic singularity in the past of any model of the Universe, in agreement with general relativity, and containing the total amount of observable matter

A singularity is considered to have zero volume and an infinite density.

In general relativity, singularities mark the boundary of the validity of this theory, hence the many unification theories (superstrings, quantum gravity, non-commutative geometry, etc) that claim to eliminate such singularities but that have failed (L'invention du Big Bang - Jean-Pierre Luminet).

Big Bang models prohibit considering times before $t°$ when the scale radius R $(t°)$ was zero. However, time $t°$ gives rise to problems of infinities (universe concentrated into a volume that is infinitely small, infinitely dense and with an infinitely great curvature). General relativity does not incorporate the quantum description of the microscopic world and, in particular, of phenomena that go right down to arbitrarily small distances, such as those corresponding to a singularity. "According to the Big Bang models, the reconstruction of the past evolution of the scale factor of the Universe...leads to the tiny value of 10^{-35} m. The corresponding time in cosmic history is called the 'Planck era'. It corresponds to a Planck time t just after t° (10^{-43} seconds after t°). Values for temperature and density were huge, respectively 10^{32} K and 10^{94} g/cm^3.... Our physics does not enable us to go any further back into the past history of the Universe, to t°, in other words to the singularity. The validity of our reconstruction of cosmic history only covers the period between today and the Planck time, t." (Jean-Pierre Luminet).

Critiques:

The problem of the Big Bang singularity, which defeats the laws of physics, together with its insurmountable problems, is thus skated over in Big Bang models. Faced with the insurmountable difficulties raised by the singularity problem, the supporters of the Big Bang can only conceal the problem, just as they do for the origin of the Big Bang. In fact, this renunciation corresponds to an admission of defeat for the theory and to the recognition of a reality which is totally inexplicable by the standard model of cosmology, the Big Bang model.

4) The problem of the homogeneous, isotropic Universe

According to the standard Big Bang model:

Observations show that the Universe is homogeneous and isotropic. The COBE satellite, launched in 1989, confirmed that the temperature of the cosmic microwave background (approximately 2.73 kelvins) is isotropic, i.e. identical in all directions, varying by less than one part in a hundred thousand.

It can be shown, using the Friedmann equations, that a homogeneous and isotropic universe will remain in this state. However, it is hard to argue that this homogeneous and isotropic state of the Universe was originally thus so. There is no evidence or valid reason to assume the existence of a homogeneous and isotropic Universe right from its origin. There is also no valid explanation for anisotropy in the cosmic background radiation of around one part in a hundred thousand.

The solution is the same paradigm that provides a solution to the horizon problem: inflation. The parts of the Universe that are observable today were causally linked before inflation. After inflation, the size of the Universe had increased by 10^{50}, and the result is homogeneous and isotropic radiation in every region of the Universe.

Critiques:

The solution to the problem of the homogeneous, isotropic Universe is similar to that to the problem of flatness and critical density, and is therefore subject to the same critique: there is no evidence or valid reason to assume the existence of a homogeneous and isotropic Universe right from its origin. There is also no valid explanation for anisotropy in the cosmic background radiation of around one part in a hundred thousand.

1) As for the horizon problem, the supporters of the Big Bang model are forced to admit: "there is no evidence or valid reason to assume the existence of a homogeneous and isotropic Universe right from its origin. There is also no valid explanation for anisotropy in the cosmic background radiation of around one part in a hundred thousand."

2) Inflationary theories, which are highly speculative, are supposed to solve the problem of the homogeneous, isotropic Universe even though they are themselves a source of serious difficulties. The solution proposed for the flatness problem, which is identical to the solution to the problem of the homogeneous, isotropic Universe, namely inflation, therefore suffers from the same difficulties, in other words it is a highly speculative hypothesis, with the *ad hoc* concept of matter with negative pressure, lacking any observational backing, and with a totally unacceptable value, 60 to 120 times greater than the value inferred from cosmological observations.

3) We should recall the critical opinion of James Peebles about inflationary theory, which we quoted earlier, and which also applies to this problem (see "Inflationary theories", Peebles p 87).

- :-

The critical analysis of the many problems raised by the standard Big Bang model (the horizon problem, the problem of flatness and critical density, the problem of the homogeneous, isotropic Universe, etc) leads us to see that the only possible solution to these problems is a highly controversial hypothesis , the concept of inflation, as we saw earlier. Far from being a solution to these problems, the use of the *ad hoc* concept of inflation does nothing more than pile up uncertainties on top of other uncertainties (new epicycles on top of other epicycles!)

The Hubble constant, Ho – The age of the Universe, to

According to the standard Big Bang model:

Hubble discovered the redshift of distant galaxies by observing a type of variable star called Cepheids. By observing the variation in the brightness of Cepheids, whose period is related to their absolute luminosity, it is possible to calculate the distance of celestial objects. The redshift was interpreted as being due to their recession velocity and to the Doppler-Fizeau effect. The linear relationship between redshift and the distance of galaxies was discovered by Hubble in 1929. The recession speed of galaxies is proportional to their distance. The constant of proportionality is called the Hubble constant, and written Ho. Hubble's law is expressed simply as: $v = H_o \times d$ where v = the recession velocity in km/s, H_o = the Hubble constant in km/s/Mpc, and d = the distance in Mpc.

Today, Hubble's law is interpreted not as being caused by the motion of galaxies through space, but rather by the expansion of space itself (within the framework of general relativity rather than special relativity, since the latter prohibits speeds faster than the speed of light, c). In 1929, the value of the Hubble constant was estimated to be around 500 km/s/Mpc, due to a wrong estimation of the absolute magnitude of the Cepheids. Today, the value of Ho is estimated to be in the region of 70-71 km/s/Mpc. In fact, the recession velocity is not constant, since the expansion of space slowed down for several billion years, and has now been increasing for approximately five billion years, with the two effects more or less cancelling each other out.

The age of the Universe represents the time that has passed since the Big Bang, i.e. the dense, hot phase of the Universe.

If the acceleration of the recession velocity of galaxies is constant, it can be found using many different methods: Cepheids, type Ia and type IIa supernovae, the study of the fundamental plane of the galaxies, and shifts in fluctuations of brightness of the multiple images of quasars produced by gravitational lensing effects. The age of the Universe, to = 1 / Ho if the Universe has very low matter density, which is what is shown by observations (an almost flat Universe).

Corrections can be made to Hubble's law.

General relativity and the Friedmann-Lemaître equations lead to a change in the scale factor R(t) as a function of time t, the expansion of space implying that R(t) becomes greater.

Observations since the 1920s show that the time periods observed (especially the periods of spectral lines) increase when more distant sources are observed. It is inferred from this that the end of the signal had to travel over a larger distance than did the beginning of the signal, and that we are therefore in a phase where the scale factor R (t) is increasing.

The expansion rate H (t) is defined as the logarithmic derivative of R (t). Its reciprocal would therefore be the time that has elapsed since a singularity at R = 0 if the increase in R (t) is linear. In 1927, Georges Lemaître provided the first theoretical expression of H (t), estimating by observation its present value Ho to be 2×10^{-17} s (or 625 km/s/Mpc), that is, 1 / Ho = 1.6 $\times 10^9$ years. This first estimation of Ho (which was named 'Hubble constant' from 1929 on) was too high, and therefore the age 1 / Ho too low, due to the erroneous estimations of extragalactic distances then available.

The current expansion rate Ho is today estimated to be 10 times lower (70 km/s/Mpc), that is, 1 / Ho = 14 × 10^9 years. The theory's other free parameters (the density parameter for the Universe and the cosmological constant) have begun to be determined observationally since 1998. They cancel each other out, giving an age close to 1 / Ho. In 2008, the value of to in the 'concordance' model was estimated to be between 13.7 and 13.8 billion years.

The distance, d, is not directly accessible. What is obtained is the distance d L (luminosity distance) or the distance, d A, (angular distance). Redshift is therefore used.

The age of the Universe can be estimated using methods that are independent of the Hubble constant: star clusters (with an accuracy of 10%), half-lives of atomic nuclei (uranium-235 – one billion years, uranium-238 – 6 billion years).

The results are as follows: globular clusters, 12-16 Ga, or 11-18 Ga (uncertainty due to the lack of precision about the distance of the clusters and about the fine details of stellar evolution); individual stars: very old white dwarfs observed with the space telescope, 12-13 Ga; the oldest stars observed, 12-16 Ga; atomic nuclei: estimations are inaccurate and give, for nuclei of uranium, thorium, osmium and rhenium, ages ranging from 10-17 Ga.

These various (astrophysical and atomic) estimations of the age of the Universe have similar orders of magnitude, around 14 ± 3-4 Ga.

The latest data provided by WMAP 5 (Table 7 – Cosmological Parameter Summary – 2008) gives a value for Ho = 71.9 (+2.6 – 2.7) km/s/Mpc and to = 13.69 (± 0.13) billion years. These values are approximately in line with (astrophysical and atomic) estimations of the age of the Universe (14 ± 3-4 Ga).

Critiques:

1) Estimating the age of the Universe by studying its components (stars, globular clusters, galaxies, atomic nuclei) gives orders of magnitude that are far too great, ranging from 11 to 18 billion years. This information can

only be useful if it is possible to establish a narrower and more accurate range.

2) The latest values for the Hubble constant, Ho, and for the age of the Universe to cited above are the result of 80 years of observational research and successive approximations. Over the decades, the value for Ho has changed from 625 km/s/Mpc to 71.9 km/s/Mpc (+ 2.6 – 2.7) and for to from 1.6 billion years to 13.69 (± 0.13) billion years. In his temporalistic model, the author established, theoretically, in 1962, a value for the Hubble constant, Ho, of 67.71 km/s/Mpc and for to (which he called the 'temporalistic constant, To') of 4.5546×10^{17} s, that is, around 14.43 billion years.

Comparing the observational value and the theoretical value for Ho, 69.2 km/s/Mpc for the former and 67.71 km/s/Mpc for the latter, there is a difference of 2.16%. This difference is negligible if we consider the uncertainty in the WMAP 5 data: between 3.2% (+2.6) and 3.75% (-2.7).

3) The value of Ho provided by WMAP 5 was obtained after decades of research and corrections, of which 69.2 km/s/Mpc is the most recent but certainly not the final result, whereas the theoretical value proposed by the author dates from 1962 and has never changed.

4) The value for the Hubble Constant, Ho, provided by NASA is the result of a great many cosmological observations and the unremitting work of a vast number of researchers. However, due to the very nature of the observations, the accuracy of the results can only be relative (like, for instance, the distance of distant celestial bodies such as galaxies or galaxy clusters, whereas the value of the Ho constant theoretically established and proposed by the author is very accurate since it is based on the values of the universal and/or quantum constants that he uses as well as on their accuracy (c, G, h, e).

From Hubble's Law, v = Ho × d, where v = recession speed in km/s, Ho = Hubble's constant in km/s/Mpc and d = distance in Mpc, we get Ho = v / d = 69.2 km/s / 3.084×10^{19} km (3.15576×10^7 s × 10^6 × 3.26 × 2.997925×10^5 km/s) = 2.243×10^{-18} s. If the Universe has a very low matter density, which is the case, the age of the Universe, to, equals 1 / Ho = 1 / 2.243×10^{18} s = 4.458×10^{17} s, which is around 14.12 billion years. The differences with the values obtained by the author are, as for the values of Ho, in the region of 2.15% (Ho = 67.71 km/s/Mpc and To = 4.5546×10^{17} s), in other words within the range of uncertainties.

In conclusion, the temporalistic model refutes the interpretation of the origin of redshifts as being the expansion of space, and interprets redshifts (the increase in wavelength of moving photons) as being physical phenomena caused by the existence of the temporalistic constant, To, with a value of 4.5546×10^{17} s. According to quantum theory, such redshifts mean that there is a loss of energy in photons as they move through spacetime. Redshifts do not have a <u>spatial</u> meaning (the expansion of space in the Big Bang model), but rather a temporalistic meaning (<u>temporal</u> effect of the temporalistic constant To on the energy of photons in motion). In other words, the cause of redshifts is of a <u>temporal</u> nature rather than a <u>spatial</u> nature. Redshifts result from the nature of photons, which are affected, as they travel through space, by the existence of the 'temporalistic constant' To whose value is 4.5546×10^{17} s. This alteration of the energy of photons is in no way connected with the concepts of 'tired light' or of interaction with other physical particles (such as the Compton effect). See Chapter VII – Paragraph 3: the redshift z and the theoretical prediction of the Hubble constant, Ho.

The Big Bang model results, in all its different aspects, from the spatial interpretation of the origin of redshifts, i.e. the spatial expansion of the Universe. All the concepts and all the difficulties that we have analyzed are caused by the paradigm of the expansion of space, which leads to highly speculative hypotheses (inflationary theories, the multiverse, singularities, etc), which violate the laws of physics and of logic (*ex nihilo* creation of mass-energy, primordial explosion at the origin of spacetime, curvature of space, i.e. of nothingness, which is an oxymoron, etc). The Big Bang model had an unfortunate consequence, namely the increasing number of speculations by cosmologists, together with a clear rejection of the usual rigorous rules of science, Popper's 'falsifiability' and Einstein's 'observable facts'.

The temporalistic model is in complete conflict with the concepts and methodologies of the Big Bang model. It results from a single hypothesis, the existence of the 'temporalistic constant', To, from which it draws all its conclusions, strictly respecting the requirements of Popper's '<u>falsifiability</u>' and Einstein's '<u>observable facts</u>' as well as the '<u>ananthropic concepts</u>' of the temporalistic model. Because it requires a rigorous approach, this model sidesteps all the difficulties of the Big Bang model (such as singularities, *ex nihilo* creation of mass-energy, etc).

In Chapter XIV, the author compares the two conflicting models, the standard Big Bang model and the temporalistic model, together with their strengths and weaknesses. It is up to the reader to decide which of the models appears to him/her to be the most scientifically valid.

Chapter IX

The evolution of galaxies – The large-scale structures of the Universe

According to the standard Big Bang model:

Most scenarios for the formation of galaxies and large-scale structures currently favor the hierarchical model, in which structures form by successive mergers of subsystems. Understanding the relationship between the distribution of dark matter and the distribution of light, i.e. of galaxies, is what is known as the bias problem. It is the subject of much current research about the formation of large-scale structures.

Nonetheless, there are doubts about the scenario of hirearchical formation of galaxies since the Big Bang. According to the statistics that have been compiled about galaxies, these only really differ in terms of their mass. The accretion of gas appears to be the main factor in the growth of galaxies (Pour la Science – N° 374 – December 2008 p 9).

According to a new scenario for galaxy formation (unlike the standard model of formation by collisions of galaxies), galaxies form from currents of cold gas (Nature 2009 - Pour La Science March 2009 – N° 377 p 11).

By using Hubble's law of expansion, the distances of fairly distant galaxies are well known.

However, a certain number of problems are recognized by the supporters of the Big Bang model:

Why do we see some spiral galaxies, which are highly evolved structures, just a few billion years after the Big Bang (Combes)?

"Theory says that elliptical galaxies could only have formed fairly recently. However, observation reveals elliptical galaxies that are already very old. Where is the mistake?" (James Peebles – Le Big Bang – La Recherche N° 35 – Quarterly - May 2009 p. 9)

A large number of galaxies between 12.7 and 13.3 billion years old have been found.

The astrophysical ages of the oldest stars observed to date appear to lie in the range 12 to 16 Ga. Astrophysics shows that there is a cutoff point at about 14 Ga for the age of stars.

Dark matter is necessary in order to ensure the gravitational cohesion of stars in galaxies and in larger structures such as galaxy clusters. What is the nature of this dark matter? See Chapter X.

According to the cosmological principle, the Universe is homogeneous and isotropic. And yet the Universe does not seem to be in the least uniform either on small or large scales. The Universe appears to be made up of filaments where clusters, super clusters and hyperclusters of galaxies, and large-scale structures such as walls and great voids, are collected together (Rudnick 2007).

In the latest SDSS (Sloan Digital Sky Survey) survey, the great wall is 1370 Mpc across, and as we move ever further away from the Earth, the observable Universe does not become homogeneous. For instance, the Great Attractor was long thought to be a cluster hidden by the Galactic disk. The supporters of the standard Big bang model reply: "The observation of the cosmic microwave background, and its very great homogeneity and isotropy, (show) that the Universe must become homogeneous from a certain time and a certain scale onwards."

According to the Jeans instability criterion, in the absence of expansion, any mass that exceeds a critical mass collapses under the effect of its own gravity, and this collapse is exponential and very rapid.

According to the Big bang model, what do the fluctuations in the cosmic microwave background represent? They are in fact the record of the fluctuations that gave rise to the galaxies and to large-scale structures. The recombination of matter took place about 380 000 years after the Big Bang. According to WMAP5, the concordance model shows that: the age of the Universe is 13.7 Ga; it is made up of around 70% dark energy, and 30% matter, of which 5% is ordinary (baryonic) matter and 25% dark matter. The model which best fits observations is the Lambda-Cold Dark Matter model (ΛCDM).

Goods 850-05, a very faint galaxy located 12 Gly away, is forming new stars at an incredible rate (4 000 stars per year), which is a thousand times faster than the current rate of star formation in the Milky Way. There is likely to have been production of large quantities of dust very early on in the history of the Universe, and this was probably caused by the first supernovae and quasars.

When simulations of the formation of structures in a ΛCDM dark matter universe are compared with observations, three unresolved problems arise: 1) the radial distribution of dark matter in galaxies does not correspond to that inferred from the galaxies' rotation curve; one possible solution is to modify the law of Newtonian dynamics at low accelerations (Milgrom 1984); 2) "At equilibrium, the disks of spiral galaxies in simulations are ten times too small compared to observations"; 3) "the ΛCDM model predicts that all spiral galaxies like the Milky way should be surrounded by at least 400 satellite galaxies, or 400 small dwarf galaxies." But according to observations, there are at most a mere dozen or so dwarf companions. What are the solutions?" (Grandes structures de l'univers - Françoise Combes – Astronomie, May 2005).

Critiques:

The evolution of the Universe between the Big Bang and the mass-energy phase has been worked out thanks to an alliance between astrophysics and quantum field theory. The standard Big Bang model says nothing, or conceals, the period between the singularity of the 'primordial explosion' and the Planck time (10^{-43} s), characterized by physical conditions—temperatures, energy densities, zero volume—that were totally extraordinary and at odds with all the current laws of physics. This is, in fact, a concept of the *ex nihilo* creation of the Universe, leading to the *ad hoc* creation of spacetime and mass-energy. This is pure speculation without any possible scientific validation.

The model of creation and evolution of galaxies and large-scale structures in the Big Bang model raises a great number of problems: what happened before the Planck time (10^{-43} seconds)? What was the process of creation of matter? From nothing? How? What was the cause of the Big Bang? The redshift of distant galaxies revealed by Hubble, on which the standard model of cosmology is based, implies a singularity with temperature, density and energy parameters with exceptionally high values. This singularity cannot be incorporated into today's physics, since the equations of both general relativity and of quantum field theory can no longer be used due to the appearance of many infinite terms (see Chapter VIII: the Origin of the Big Bang).

From the Planck time on, the standard Big Bang model has made use of the high level of knowledge and research in high energy physics and various other disciplines of physics, such as quantum physics, nuclear physics, particle physics, etc, to work out the consequences of the premises provided by this model. On the basis of the Big Bang model scenario of the fall in the temperature of the Universe over time caused by the expansion of space, quantum physics has worked out a history of the Universe where, as temperatures, i.e. energies, decreased, the four fundamental forces (strong, weak, electromagnetic and gravitational), which were originally unified, then decoupled. As a result of the uncertainty principle in quantum physics, very short-lived virtual particles and antiparticles appeared, progressively followed by other particles. At 10^{-35} seconds, the Universe was filled with a large number of particles, including electrons, neutrinos and quarks.

Between 10^{-35} and 10^{-32} seconds inflation took place. At 10^{-6} seconds, quarks became confined, and then at around 10 seconds primordial nucleosynthesis took place. Finally, recombination took place when the temperature of the Universe had fallen to 3 000 kelvins, 380 000 years after the Big Bang. Galaxies were formed after several hundred million years.

The history of the Universe as calculated and described by quantum physics, which we have briefly summarized, cannot be refuted technically. It is based on knowledge built up over decades by quantum field theory. However, in the final analysis, its validity is only based on the premises provided by the Big Bang model, in other words the interpretation of the redshifts of distant galaxies as being caused by the expansion of space. If this interpretation is refuted, the entire history of the Universe that is inferred from it becomes totally invalidated.

The energy fluctuations that arose several thousand years after the Big Bang, from which galaxies are supposed to have formed through the action of gravity, are not enough to justify the evolution of large-scale structures. According to Tegmark(2004), although the anisotropies in the cosmic microwave background completely match on small and medium scales, this is not at all true on large scales. The way in which the structures develop depends on the origin of the primordial fluctuations and on the nature of dark matter.

In 2004, Brigitte Rocca revealed the existence of very young massive galaxies (at distances > 12 Gly), which contradicts the hierarchical growth model (Dossier La Recherche 393 – January 2006).

Depending on the authors, the Universe has a structure similar to foam, sponge, sheets, pancakes or a three-dimensional spider's web. To sum up, it may be considered that the large-scale structures of the Universe are made up of filaments made of gas, dust, stars, galaxies, galaxy clusters and superclusters, great walls, great voids and dark matter. These filaments may represent around 10% of space and contain 15% of the galaxies and galaxy clusters. Their typical length ranges from 50 to 80 Mpc (1.5 to 2.4 × 10^{26} cm). They border on huge voids that have diameters typically ranging from 25 Mpc (8 × 10^{25} cm) to 125 Mpc (4 × 10^{26} cm). Located between 6 and 10 billion ly from Earth, the biggest void ever discovered, lying in the direction of the constellation Eridanus and discovered by Lawrence Rudnick (August 2007), probably has a diameter of around 1 billion ly (1 × 10^{27} cm). The reality of this void is disputed, due to bias in the statistics of the galaxies that have been inventoried, and we will therefore have to wait for new observations concerning it. This great void, the probability of whose existence is estimated to be 5 × 10^{-10}, as well as the various other inhomogeneous structures already discovered, seriously challenges the standard model of cosmology, based on the cosmological principle which gives the Universe a homogeneous and isotropic structure. The expanding Universe Big Bang model is forced to admit the existence of this repetitive yet irregular large-scale structure of the Universe, and especially of these huge voids measuring roughly 1 × 10^{26} cm to 1× 10^{27} cm across. The standard model is unable to explain the causes of the existence of these huge voids, whose probability of existing is tiny (5 × 10^{-10}).

On the other hand, the temporalistic model offers a simple explanation for the structure of the Universe and for the existence of filaments and great voids. In the temporalistic model, gravitation has a finite range, embodied by the concept of gravitational radius r = $m^{1/2}$ (r = radius, m = mass). In the filaments, the gravitational effect of galaxies and galaxy clusters operates lengthways, since the masses are relatively close and therefore below the threshold of the gravitational radii. If we take the example of a rich galaxy cluster (3 000 galaxies) whose mean mass is around 1 × 10^{49} g, its gravitational radius is (1 × 10^{49})$^{1/2}$ cm = 3 × 10^{24} cm. It can therefore have a gravitational influence on galaxies and galaxy clusters whose average distance is 1 Mpc (3 × 10^{24} cm) (See Chapter XII –Temporalistic gravitation – Masses and gravitational radius, paragraph 10), all along the filaments.

As for voids, galaxies and galaxy clusters and superclusters can only exert a gravitational influence at their center if their gravitational radius is equal to or exceeds the radii of the neighboring voids. For example, for a void of 1 × 10^{25} cm, the gravitational mass necessary is 1 × 10^{50} g, which is the average mass of 40 000 galaxies; for a void of 1×10^{26} cm, the gravitational mass necessary is 1×10^{52} g, i.e. the average mass of 4 million galaxies, and for a void of 1×10^{27} cm (Rudnick's void), the gravitational mass necessary is 1×10^{54} g, i.e. the mass of 400 million galaxies. The huge sizes of the masses necessary for galaxies and galaxy clusters to have a gravitational effect on great voids, and the scarcity of such concentrations of galaxies, explains the existence of these voids, which is one of the serious challenges to the Big Bang model.

Chapter X

Dark matter –The Pioneer effect – The MOND theory – The Casimir effect

The dark matter problem

According to the standard Big Bang model:

Dark matter (or missing matter) is estimated to make up around 80-90 % of all matter. It reveals its presence not only in galaxies but also in the large-scale structures of the Universe, and in galaxy clusters and superclusters. Many candidates have been proposed (MACHOs, neutrinos, WIMPs, brown dwarfs, supermassive black holes, etc) but, for now, its nature remains unknown.

What is currently known about the nature of dark matter?

1) The mass/luminosity relation according to distance confirms the existence of an invisible type of matter, not only around galaxies but also between them.

2) The rotation curve (velocity) of galaxies leads to the conclusion that stars and other luminous bodies make up less than 10% of the total mass of a galaxy. The remaining 90% is made up of dark matter or is under the influence of an unknown phenomenon.

3) The rotation curve of galaxies suggests that dark matter is contained in vast halos surrounding the galaxy's visible stars.

4) It is impossible to find dark matter far from galaxies, in very extensive halos, because tidal forces would disperse it throughout the entire cluster containing the galaxies.

5) Studying the effects of dark matter using the gravitational lensing method on the galaxy cluster Abell 1689 (distortion depending on the mass and radius of the deflecting galaxies), as proposed by the physicist Anthony Tyson, shows that "dark matter makes up over 90% of all matter".

6) To a large extent, dark matter accompanies luminous matter wherever it is located in galaxies, galaxy clusters and even large-scale structures tens of megaparsecs across.

7) Dark matter follows the irregularities in the density of distribution of luminous matter throughout the visible Universe.

8) Dark matter does not exist or does not reveal its presence in the great voids that are tens or hundreds of megaparsecs in size (Richard Schaeffer 2001- see Chapter XI, paragraph 4, of the temporalistic model).

According to the temporalistic model:

We hypothesize that the temporalistic acceleration field is the same thing as dark matter. Below, we detail the arguments that back up this proposition:

1) Since the temporalistic field emanates from photons, therefore from luminous sources, it fits the spatial distribution of dark matter.

2) Because of its origin (luminous bodies), the temporalistic field necessarily follows the irregularities in the density of distribution of luminous matter throughout the visible Universe.

3) According to the temporalistic model, the temporalistic field results from the gradual dampening down of photons' vibrations and therefore from a constant loss of energy (redshift), caused by the existence of the temporalistic constant To. The fact that 90% of dark matter in the Universe is located in the neighborhood of luminous sources agrees well with the hypothesis that it has a temporalistic origin.

4) Conversely, since great voids do not contain luminous matter, they cannot therefore contain dark matter.

5) the temporalistic field is not a hypothetical field but a field which necessarily results from the temporalistic model. The acceleration caused by dark matter is the consequence of the existence of the temporalistic gravitational constant G', whose value is 6.582×10^{-8} cm/s^2.

6) The temporalistic field, whose vectors are gravitons, is not a luminous field.

7) The paragraph below concerning the anomalous radial acceleration of Pioneer 10 and other spacecraft demonstrates the existence of the

temporalistic universal isotropic acceleration field G' = 6.582 × 10⁻⁸ cm/s²
and validates it.

8) The value of the acceleration of the speed of stars in galaxies, attributed to the effect of dark matter, is of the same order as the value of the temporalistic gravitational constant G', that is, 6.582×10^{-8} cm/s². Its value is also of the same order as the modification of Newtonian gravitation proposed by the MOND theory (which rejects the existence of dark matter).

9) Dark matter does not feel the nuclear force, the weak force or the electromagnetic force. It only feels gravitational force, which agrees well with the fifth proposition of the temporalistic field: The acceleration of dark matter is the result of the temporalistic gravitational constant G', that is, 6.582×10^{-8} cm/s².

10) Very recent observations Benoit Famaey and colleagues – Strasbourg Observatory - G. Gentile et al. Nature, 461, 627-628, 2009) clearly confirm the correlation between luminous matter and dark matter: "Astonishing relationships appeared: from one galaxy to another, the strength of gravity caused by dark matter at the characteristic radius is identical, and the same applies to the strength of gravity caused by visible matter at the same radius. What can be inferred from these relationships? Firstly, that there is an inverse correlation between the central density of dark matter and that of visible matter. A high central density of visible matter implies that the density of dark matter at the center is low, and vice versa. Secondly, that the ratio of the densities of visible matter and dark matter which applies over the whole Universe remains valid within the characteristic radius for all galaxies."

These observations are consistent with the temporalistic model. According to this model, dark matter results from luminous sources. The various paragraphs above which set out the temporalistic model for dark matter corroborate, both qualitatively and quantitatively, the observations in paragraph 10.

The anomalous radial acceleration of Pioneer 10

According to the standard Big Bang model:

For over 20 years, a problem has intrigued planetary scientists and physicists: "a slight, unexplained sunward acceleration of the motion of the

Pioneer 10, Pioneer 11 and Ulysses spacecraft (www.geocities.com/solarstormmonitor/Pioneer.html). Many other websites provide information on this subject.

The anomalous acceleration has several characteristics:
1) Its value, according to different authors, is 7.59×10^{-8} cm/s^2 (http://renshaw.teleinc.com/papers/prl-pi/prl-pi.stm), $8.74\ (\pm\ 1.33) \times 10^{-8}$ cm/s^2 (http://csep10.phys.utk.edu/newsgroups/mond/messages/22.html), or "Around 10 billion times smaller than the acceleration we feel, the Earth's gravitational attraction" (www.geocities.com/solarstormmonitor/Pioneer.html - http://spaceprojects.arc.nasa.gov/Space_Projects/pioneer/PNStat.html).

2) The order of magnitude of this anomalous acceleration is c × Ho (Hubble constant).

3) This anomalous acceleration, which is independent of distance, is constant with regard to the speed of the spacecraft.

4) This anomalous acceleration is radial.

According to the temporalistic model:

1) This unexplained effect results very precisely from the existence of the temporalistic universal isotropic acceleration field G' = c / To, where G' is the temporalistic gravitational constant, c the speed of light, and To the temporalistic constant, that is 6.582×10^{-8} cm/s^2 = 2.997925×10^{10} cm/s / 4.5546×10^{17} s.

2) The order of magnitude of this anomalous acceleration c × Ho (Hubble constant) corresponds to the temporalistic model with c / To (Ho = 1/To) = G ' (temporalistic gravitational constant).

3) When spacecraft leave a circular or elliptical path and take on a radial path leading out of the Solar System, the radial effect of the temporalistic universal acceleration field makes itself felt, and reduces the speed of the spacecraft (Pioneer 10, Pioneer 11, Ulysses, Galileo, etc.)

4) The temporalistic universal acceleration field does not affect the circular or elliptical orbits of the planets in the Solar System but only radial paths.

5) An experimental measurement validates the temporalistic model. By September 1998, the slowing down of Pioneer 10 had caused it be some 400

000 km closer to the Sun compared to its predicted path. The radial journey of Pioneer 1, which began in 1973 – 1974, had therefore lasted some 24.5 years, i.e. 7.73×10^8 s. During this period, the deceleration, with an acceleration constant of 6.582×10^{-8} cm/s^2, equalled 6.582×10^{-8} cm/s$^2 \times 7.73 \times 10^8$ s $\times 7.73 \times 10^8$ s $= 3.93293 \times 10^{10}$ cm = <u>393 293 km.</u>

<u>The MOND theory</u>

http://nedwww.ipac.caltech.edu.level5/Sept01/Milgrom/Milgrom_contents.html
121

The MOND theory proposes that when the acceleration inferred from the Newtonian acceleration constant Gn is less than a°, that is Gn < a°, Newtonian theory does not apply, the parameter a° being comparable in value to c × Ho. According to the temporalistic model where Ho = 1 /To, a° ~ c × Ho, which is 6.582×10^{-8} cm/s^2.

The MOND theory is proposed as an alternative to dark matter. The temporalistic model does not deny the existence of dark matter. When the acceleration caused by a mass is smaller than G', the Newtonian model no longer applies in MOND theory. In the temporalistic model, Newtonian theory no longer applies for an acceleration smaller than G', as in MOND theory, but this is due to the finite gravitational radius of masses and to the <u>temporalistic universal acceleration field G'</u>. (See Chapter XI).

The Big Bang model does not support MOND theory.

<u>The Casimir effect</u>

<u>According to the standard Big Bang model:</u>

The Casimir effect, named after its discoverer, is an effect that exists between two parallel conducting metal plates which, when placed very close to each other, attract each other. This force is supposed to result from the concept of the quantum vacuum, which is not really a vacuum but the seat of fluctuations that create virtual particles that exert an attractive force on the plates.

<u>According to the temporalistic model:</u>

The temporalistic model proposes an <u>alternative to the quantum explanation.</u>

The isotropic acceleration field of value G' created by the photons' loss of energy is disturbed by the presence of the two metal plates. The result is that there is a smaller acceleration force between the two plates than on the outside of the plates, causing them to move closer to each other. It would be interesting to calculate whether this temporalistic effect is quantitatively confirmed.

PART FIVE

(See CALCULATIONS –CHAPTER XV – page 181)

AN ALTERNATIVE TO THE BIG BANG MODEL

Chapter XI

The temporalistic model

The concept of time and the constant To

In Chapter III: Anthropic and ananthropic concepts (b: the physical concept of time), we analyzed the concept of time.

Below, we summarize the results of this analysis.

In the theory of special relativity, Einstein (1905) introduced a new concept, that of an inseparable space and time, four-dimensional spacetime. From this perspective, time appears as a fourth spatial dimension directed from the past towards the future, in this way defining a 'light cone'. Time and space, which are intimately linked, constitute frames of reference against which physical phenomena, such as momentum, energy, speed, etc are measured. Physical laws are invariant with regard to a change in the reference frame. Quantum physics, which integrated special relativity into quantum electrodynamics, has hardly altered the relativistic concept of time. It even made it more radical, in a spatial sense, in Feynman diagrams, where the direction past > future is no longer preferred over the direction future > past (particles and antiparticles). By correlating uncertainty about energy with uncertainty about time, Heisenberg's uncertainty relations also give no specific definition of time. Although Einsteinian relativity emphasizes the arrow of time past > future (the light cone), it abolishes the notion of time for photons. For a moving clock, time slows down. For a clock traveling at the speed of light, time would slow down infinitely.

A photon traveling through a vacuum at the constant speed c is, according to Einsteinian relativity, unchanging and is therefore located outside time.

In superstring theories, the Universe is made up of eleven dimensions, including seven spatial dimensions wrapped up in Calabi-Yau spaces, and four visible dimensions in spacetime. In the time dimension, the photon does not age. "<u>At the speed of light, time ceases to flow</u>" (Brian Greene 2000).

In the final anlysis, time is conceived of as a fourth spatial dimension of the Universe. The direction past > future disappears for photons. Past > future asymmetry is the only parameter that distinguishes the spatial dimensions from the temporal dimension. This asymmetry, refuted by Stephen W. Hawking, is asserted by Roger Penrose (1996). If asymmetry disappears from the concept of time, there is no longer anything to distinguish the temporal dimension from a spatial dimension. A recent experiment has confirmed the asymmetry of time in strange elementary particles (PLEAR 1998).

The temporalistic model results from the the hypothesis of <u>the fundamental asymmetry of time</u>: < http://site.voila.fr/nobigbang> (Chapter 5: The concept of time).

The temporalistic hypothesis

The temporalistic hypothesis is based on the fundamental asymmetry of time. According to the temporalistic hypothesis, the asymmetry of time is an integral part of physical phenomena, including the nature of the photon. According to current physics (Einsteinian relativity, quantum mechanics, superstring theory), when a photon is emitted by an atom in a star that is distant in space, and therefore in time, then as long as no external interaction occurs, if the photon travels through the vacuum to an Earth-based telescope, the photon will be observed exactly as it was emitted, whether two million or two billion years earlier, that is, with the same energy $w = h\nu$ (h = the Planck constant; ν = frequency), the same momentum $p = h\nu/c$, and the same wavelength $\lambda = c/\nu$. None of the measurable quantities of the photon have changed. It travels through space unchanged. These are the facts, or rather <u>the assumptions of current physics </u>(naturally without taking into account the expanding Universe hypothesis).

Special relativity postulates that light (photons) travels, in a vacuum, at a constant speed, c, without any alteration of these parameters. On the other hand, the temporalistic model proposes that the oscillations of light waves are gradually dampened down as they travel through a vacuum.

For the temporalistic model, the characteristics of emitted photons are modified as they travel, without any external interaction on the photons. In other words, the model incorporates the concept of time into the very nature of the photon. The redshift of distant galaxies is currently interpreted as being a cosmological effect caused by the expansion of the Universe. The temporalistic model recognizes that redshift is a fact. However, it does not interpret it as resulting from a known physical effect (Doppler effect, cosmological effect, Compton effect, gravitational effect, etc). It is interpreted as being an intrinsic part of the physics of the photon. It is considered as being a structural, quantum property of the photon, caused by the existence of an unrecognized temporal parameter, which is given the name 'temporalistic constant, To'. It is this constant which affects the photon and which is the stamp of the asymmetry of time. We shall now see how we can go about searching for the constant, To, postulated by the temporalistic model.

The search for the constant To

The search for the temporalistic constant To can follow various routes. Theoretical considerations about the structure of the Universe can help us, as well as dimensional analysis. The first thing that we can observe is that this parameter does not clearly appear in quantum phenomena, since it has never been detected. We can therefore say that, if it exists, it is either hidden or unrecognized.

What are the major physical constants that might help us in the search for this unrecognized temporalistic parameter? We have selected four: c, h, e and G.

These different physical constants appear as boundaries in our physical Universe: the upper limit for speed (c); the lower limit for actions (h ; the lower limit for electric charge (e, the elementary electric charge, is the lowest free charge known, since the fractional charges of quarks and antiquarks concern confined particles). The status of G can also be considered as a yardstick for the strength of interaction exerted by one mass on another mass (Newton) or as that for the strength of interaction exerted between masses and energy, on the one hand, and the metric field on the other (Einstein). Just like these fundamental physical constants which make up the limits of our physical universe, the temporalistic model conceives of the temporalistic constant To as being another of Nature's limits: in addition to the limits on speed, actions, electric charge and gravitational interactions, there is a limit on time. We define this time limit as being a boundary for time, just as c is a boundary for speed, h a boundary for actions, etc.

We call this limit on time the temporalistic constant or To constant. According to the temporalistic model, the motion of a photon through space, that is, through time, results in a damping down of its oscillations, i.e. by a change in its parameters (energy, wavelength, frequency, etc), which becomes apparent as redshift. Let us attempt to specify the dimension and numerical value of this temporalistic constant To. At first sight, it seems logical to assign a dimension of time to this constant, since this is how the temporalistic hypothesis postulates the origin of redshift in photons. We shall see by the results obtained that this assumption is a valid one. The search for a temporalistic constant with the dimensions of time can be undertaken using the dimensional criteria To = L / LT^{-1} (length / speed), or LT^{-1} / LT^{-2} (speed / acceleration), or MLT^{-1}/MLT^{-2} (momentum / force), or ML^{2}T^{-1} / ML^{2}T^{-2} (action / énergie), or using more complex formulae such as (h × G / c ×10^5)^{½} (angular momentum × gravitational constant / speed of light × 10^5)$^{-1/2}$. However, a purely dimensional analysis is unable to show us the route to follow in order to find the temporalistic parameter we are seeking.

We therefore have to try another way, while respecting dimensional consistency. Of the four major physical constants that make up the boundaries of our Universe, the constants h and e only appear to concern boundaries on small scales, G on large scales, and c on both. (Energy of a photon: E = hv (frequency), or hc / λ (wavelength), or E = mc^2). Establishing relationships between these limiting constants could be a suitable avenue of approach. The constant To (quantum constant) is itself a boundary or limiting constant. Establishing a relationship between h and e does not, at first sight, look promising, while h and G, or e and G, seem scarcely better (h and e are quantum constants that apparently have no connection with the macroscopic constant G).

The ratio c / G

(See CALCULATIONS –CHAPTER XV – page 181)

Our search for the constant To leads us to examine the ratio c / G. This is the ratio between the upper speed limit c for physical phenomena and Newton's gravitational constant G (the strength of attraction between masses, or, in General relativity, the extent of the curvature of the metric field caused by masses and energy). The ratio c / G tells us the maximum time over which gravitational attraction (or the strength of the effect of mass-energy on the metric field) can act in order to reach the upper speed limit c.

In Newtonian mechanics, we have (in the CGS system): To = c/ G, that is, 2.99792×10^{10} cm/s / 6.67×10^{-8} cm^3/g s^2 = 4.494×10^{17} s g/cm^2. (1)
The value of To would be 4.494×10^{17} s if the ratio g/cm^2 was approximately equal to one.

It is now necessary to refine the numerical value obtained for the parameter To. We can assume that the constant To is a quantum constant. On the other hand, the constant G is a macroscopic constant that applies to weighable masses. The accurate measurement of G was carried out using sizeable masses (compared with atomic masses) and with material whose average density is comparable to that of iron (Cavendish's torsion balance experiment 1798). Quantum physics tells us that different atomic nuclei have binding energies of varying sizes (Aston's packing fraction), and as a result, have a mass defect. The binding energy per nucleon for nuclei containing between 30 and 120 nucleons is over 8.5 MeV. It is around 9 MeV for nuclei with a mass number in the region of 56 (Fe). Since the energy of a nucleon is less than a billion electron-volts (the energy of a proton is 938.1 MeV), Aston's packing fraction is about 1% of the mass for atomic masses close to that of iron. It is therefore necessary to adjust the 'macroscopic' value of G in relation to the masses of quantum particles without nuclear binding energy, electrons, nucleons, etc. As a first approximation, the 'quantum' value of G is therefore: G = 6.67×10^{-8} cm^3/g s^2 × 99/100 = 6.60×10^{-8} cm^3/g s^2. We obtain for G': 6.67×10^{-8} cm/s^2 × 99/100 = 6.60×10^{-8} cm/s². This gives a more accurate value for To: 2.99792×10^{10} cm/s / 6.60×10^{-8} cm/s² = 4.5423×10^{17} s, that is, approximately 14.4 billion years.

Later on we will be able to refine this value of To even further, using purely quantum constants that are more precise. The value of To was established by the author in 1962.

From this value, the author was able to predict a theoretical value for the Hubble constant Ho, as was shown in Chapter VII.

The value of the temporalistic gravitational constant G' leads to a new interpretation of gravitation with a finite range. One of the consequences of this is the gravitational radius and the quantitative values of the gravitational radii of the large-scale structures of the Universe (stars, galaxies, galaxy clusters, great walls, great voids, etc). Another result is the explanation and precise value of the anomalous radial acceleration of the spacecraft Pioneer 10, 11, Ulysses, etc. Chapters XI and XII describe the various aspects of temporalistic gravitation.

If the temporalistic hypothesis is right, the parameter To, a quantum parameter, should show up in quantum phenomena. This is what we have verified in the following chapter < http://site.voila.fr/nobigbang> .

The quantum constant G'

The presence of the temporalistic constant in various quantum phenomena (Josephson effect, photoelectric effect, fine structure constant) results from our proposition $e = h/2\mu \times To$, and hence $e / h/2\mu = To$, or $h /2\mu \times e = 1 / To$.

The route we followed to find the temporalistic constant To, that is, the ratio of the speed of light c to the (temporalistic) gravitational constant G', turns out to be a pragmatic rather than a fundamental approach.

The fundamental theoretical approach sets out from the concept of To, a limiting quantum constant analagous to h, e or c. As we have just seen, this constant, which establishes a relationship between e and h, plays a major role in quantum electrodynamics. It also makes it possible to connect the constants e, h and G' with each other, by setting down the relationship G' = c / To, a ratio between two limiting quantum constants. G' is thus shown to be a quantum quantity. In Chapter XII we shall look at its large-scale, gravitational significance.

The previous chapters have shown that, in agreement with the temporalistic hypothesis, fundamental quantum phenomena directly depend on the temporalistic constant To, both with regard to dimension and to numerical value. Having reached this stage of our argument, we have seen that we can eliminate the gravitational origin of the parameter To. Its purely quantum significance and value are no longer dependent on the gravitational constant G'. The presence of the constant To at the heart of quantum phenomena is a powerful argument in favor of the temporalistic hypothesis, and one which appears inescapable. There appears to be no other way of explaining the numerical coincidences of the ratios c / G' and e / h, which are ratios between independent constants. No other explanatory argument appears plausible.

Four quantum effects

(See CALCULATIONS page 181)

The constant To, a quantum parameter, appears in <u>4 quantum effects</u>:
1) The elementary electric charge, e: h/bar × To
2) The constant of proportionality in the Josephson effect: 2 e / h, that is 2 e / h × 2μ, and, in angular frequency, 2 To
3) The constant of proportionality of the stopping potential in the photoelectric effect equals 1 / To.
4) In the temporalistic model, the fine structure constant shows up as the ratio between the elementary electric charge e and the parameter G' (c / To).

CHAPTER XII

Temporalistic gravitation

The existence, in the physical Universe, of the constant To, has major consequences for the temporalistic approach to gravitational forces, or, more specifically, to the phenomenon of gravitation. The author does not claim to set forward a new model of gravitation here. He has simply attempted to show how the temporalistic model has of necessity implications for the interpretation of the phenomenon of gravitation.

We have seen that the existence of the constant To led, without the need for any other hypothesis, to the phenomenon of the redshift, z, of galaxies (Chapter VII). Redshift in electromagnetic radiation corresponds to a fall in the frequency v or w (angular frequency) of the photon and hence of its energy. Correlatively with redshift, to a first approximation it can be considered that the decrease in the frequency of the photon is proportional to the time, t, elapsed between the time it was emitted, Te, and the time it is received, Tr, that is Åw = E - E' / E = t / To (where E is energy emitted and E' energy received). The energy of the photon E = hw therefore varies over time, and its decrease is proportional, also to a first approximation, to the travel time of the photon ÅE = hwe - hwo / hwe = t / To (where hwe is energy emitted and hwo energy observed).

Talking about the redshift or the fall in frequency (or energy) of a wave comes to the same thing, because they are different aspects of the same quantum phenomenon. The value of this temporalistic quantum phenomenon is very low. This is the fundamental reason why quantum mechanics has not detected it or taken it into account until now. It only appeared, in the Hubble-Humason effect, at minimum distances of ½ to 1 million parsecs, that is, 1½ million to 3 million light years. The recession of galaxies due to expansion only appears beyond the Local Group, where it can be distinguished from the local speeds of galaxies. Nevertheless, the phenomenon of decreasing energy of photons as they travel through space is a continuous phenomenon that affects all photons as soon as they are emitted.

The existence of the constant To entails an evolutionary concept of the energy of photons. Just as the neutrino was 'invented' by Pauli to explain energy balance in radioactive decay, it seems to us that a rational and fruitful approach is to assume that the loss of energy by photons as they travel through the Universe takes place via the emission of particles (or waves) by the photons. There appears to be no other alternative that respects the law of conservation of energy, while at the same time allowing an evolutionary concept of the physics of photons, in agreement with the existence of the constant To. We shall see that, when this hypothesis is compared with the evidence, this assumption turns out to be justified.

Starting out, then, from the hypothesis that the decrease in energy of photons (or redshift) leads to the emission of new particles, which we shall call X particles, this results in the hypothesis of a field of X particles to which, for convenience, we shall give the name temporalistic field. It immediately becomes obvious that the existence of this temporalistic field has a considerable effect on gravitation.

Gravitation has gone through three major stages in the history of physics: Newtonian gravitation, the relativistic theory of gravitation and the metric theories that are derived from it, and attempts at formulating quantum theories of gravitation. The superstring theory appears to imply, theoretically, the existence of a gravitational field (Brian Greene 2000).

In these three groups of theories, gravitation is defined either as coupling between masses (Newton 1687), or as a curvature of the metric field by masses and energy (Einstein 1916), or as an exchange force between quantum fields whose carrier is the graviton (quantum field theory). What these different interpretations of gravitation have in common is the existence of a field (attractive, metric or quantum) whose strength is given by the coupling constant G or its Einsteinian derivative $8\mu G/c^2$. This constant has a fortuitous, empirical value which does not result from the theory. The strength of the gravitational field is observed: it is not conceptually inferred from the theory.

In these three groups of theories of gravitation, if we leave aside the sources of the field (masses and energy, or particles), space can be considered to be an empty or almost empty space, apart from quantum fluctuations. In the temporalistic model this is impossible, because photons continuously supply the temporalistic field, whose carrier is the X particle. Space has an energy level corresponding to the continuous production of X particles by photons.

How can we determine the energy state corresponding to the existence in space of the temporalistic field? This is where we see the deep significance and justification of the dimension of the parameter G' that we used earlier.

We formulated the hypothesis of a dimension LT^{-2} (that of an acceleration) for the gravitational constant G' in the temporalistic model, whereas the Newtonian dimension is $M^{-1}L^3T^{-2}$. In Newtonian theory, as in relativistic gravitation, the gravitational constant G is a parameter related to mass and energy. This parameter, as we recalled earlier, gives the strength of the coupling between masses, or between masses (and energy) and the metric field.

Temporalistic gravitation interprets the constant G' in a different way. The existence of the temporalistic field implies that of an energy field in space, even in the absence of particles of matter or of energy. In agreement with the cosmological principle, we may consider, to a first approximation, the Universe to be isotropic and homogeneous. The energy field of the temporalistic field can be thought of as the energy associated with the gravitational potential of a universal gravitational field. This gravitational potential, which is equivalent to an acceleration, is derived from the temporalistic field and no longer from the masses present. By bringing together the two limiting constants c and To we obtain its value: G' (universal acceleration constant) × To (upper limit on time) = c (upper limit on speed). In SI, 6.582×10^{-10} m/s² × 4.5546×10^{17} s = 2.997925×10^8 m/s. In the CGS system, 6.582×10^{-8} cm/s² × 4.5546×10^{17} s = 2.997925×10^{10} cm/s. In dimensions, LT^{-2} x $T = LT^{-1}$.

In temporalistic gravitation, the constant G' is therefore a constant related to the universal temporalistic field, and its dimension is that of an acceleration. G', in the temporalistic model, is no longer related to matter, which explains why M has disappeared in its dimensional equation. G' is the universal acceleration potential related to the existence of the temporalistic field. G', a ratio between two quantum constants, c and To, therefore also appears as a quantum constant. G', the constant of temporalistic gravitation, is no longer an empirical parameter, calculated from observations. It results, theoretically, from the ratio between the two constants c and To.

How does the temporalistic model interpret the phenomenon of gravitation?

In the temporalistic model, mass and energy are no longer, as in classical theories of gravitation, the sources of (gravitational or metric) fields. Mass and energy are considered to be parameters that disturb the universal acceleration potential. The carriers of this universal acceleration potential, X particles, can be thought of as being gravitons. Mass and energy, via a screening effect (scattering or absorption), disturb the balanced, isotropic temporalistic field of gravitons, whose potential, in the absence of mass, should be considered as an acceleration potential of value G'. The presence of matter and energy exerts a disymmetrical action on this acceleration potential through the screening effect that it has on the propagation of X particles, or gravitons. It is the local alteration of the acceleration potential caused by the disturbing effect of mass and energy that appears to the observer as a gravitational phenomenon (Newtonian theory) or as a curvature of spacetime (relativistic gravitation). This alteration of the isotropic acceleration field by mass and energy thus appears as a disymmetrical force field which attracts masses, or as a curvature of the four-dimensional metric field.

In the final analysis, the disturbing effect of masses on the temporalistic acceleration field can be likened to their cross section according to $M \sim L^2$. By applying this value to the dimensional equation for the pressure, we obtain $ML^{-1}T^{-2} = L^2L^{-1}T^{-2} = LT^{-2}$ (an acceleration).

The temporalistic acceleration field can be likened to a pressure field whose dimensional equation is given by p (pressure) = F (force) / S (surface area), that is, $MLT^{-2} / L^2 = L^2L^{-1}T^{-2} = LT^{-2}$ (an acceleration). The disturbance parameters, or the sources of the gravitational or metric field, are proportional to the masses. Now, the scattering cross sections of the masses are also proportional to the masses. The barn (10^{-24} cm^2) is the cross section of a large nucleus of matter (with an approximate value of 10^{-24} g). We saw in Chapter IX that the ratio g / cm^2 is approximately equal to one. We may therefore establish a principle of nuclear equivalence between cross section in cm^2 and matter in grams. Since nuclear density is roughly the same for all atomic nuclei, the cross section of atoms is, to a first aproximation, proportional to their mass: $M \sim L^2$. We must nevertheless point out that the screening effect of the disturbance parameters of masses depends on their nuclear composition, since the nuclear density of nuclei varies slightly according to their nuclear composition. The screening effect of the disturbance parameter of masses is also inversely proportional to the square of the distances of the masses $1 / r^2$.

We know that the gravitational forces or the curvature tensors of metric space are proportional to masses (and to energy). This fact can be explained logically in temporalistic gravitation since mass corresponds to the disturbance parameter of the universal isotropic gravity field. We have seen that the disturbance power of a mass can be considered to be the same, to a first approximation, as that of its cross section L^2. The mass acts by distorting the isotropic acceleration field, and this action becomes more significant the greater the mass (or the corresponding cross section L^2) and the closer the space in question is to the mass. The constant of proportionality with distance is given by the well known factor $1/r^2$.

In classical theories of gravitation, gravitational interaction has an infinite range. In the temporalistic model, this is impossible. The range of disturbance of the gravity field of gravitons by masses is limited by the value of the universal acceleration field, that is, G' (6.582 × 10^{-10} m/s² in SI or 6.582 × 10^{-8} cm/s² in the CGS system). The disturbance caused to the universal acceleration field by the presence of masses and energy becomes apparent via the emergence of a local acceleration field. This screening or disturbance effect will only be felt if it is greater than the universal acceleration field. In other words, if the strength of the disturbance to the universal acceleration field caused by the screening effect of the masses is less than G', the disturbance effect of the masses will no longer be felt. Temporalistic gravitation therefore has a <u>limited range</u>.

Let us compare the expression of the gravitational force that is exerted between masses m and m':

In Newtonian theory, $F = Gmm'/r^2$ and the dimensional equation gives $F = M^{-1}L^3T^{-2} \times M^2/L^2 = MLT^{-2}$.

In the temporalistic model, $F = G'mm'/r^2$ and the dimensional equation gives $F = LT^{-2} \times L^2 \times L^2/L^2 = L^3T^{-2}$ and, by applying $M \sim L^2$, $L^3T^{-2} = MLT^{-2}$.

In the temporalistic model, we obtain, for the terrestrial gravitational field, $g = G'M/r^2$, that is, $LT^{-2} \times L^2/L^2 = LT^{-2}$.

We can calculate the <u>limited range of temporalistic gravitation</u> by using, to a first approximation, the temporalistic dimensional equation of Newtonian gravitational force: $F = G'mm'/r^2$, that is, $F = LT^{-2} \times L^2 \times L^2/L^2 = L^3T^{-2}$ and, by applying $M \sim L^2$, $L^3T^{-2} = MLT^{-2}$. For the local acceleration field, mG'/r^2, we obtain $L^2 \times LT^{-2}/L^2 = LT^{-2}$.

In order to be felt, the local acceleration field must be greater than the universal acceleration field G'. We can therefore write $mG'/r^2 > G'$, hence $mG'/G' > r^2$ or $m > r^2$, that is $r < m^{1/2}$ and with the equivalence $M \sim L^2$, we obtain $r < L$.

For concentrations of matter in the Universe, temporalistic gravitation therefore sets an upper spatial limit given by the approximate formula <u>r = $m^{1/2}$. This is the gravitational radius of masses.</u> This restriction only applies in temporalistic gravitation. It does not apply in other theories of gravitation because in these theories gravity's range is infinite.

General relativity (geometrical) and the temporalistic model (physical)

General Relativity: a geometrical model

The concept of spacetime in general relativity

The spacetime continuum in general relativity has four dimensions, three in space and one in time; an event is located in time and space by its coordinates ct, x, y and z, which all depend on the reference frame. In the current state of knowledge, only spacetime, conceived of as a unified concept, is mathematically a Minkowski space in special relativity and any kind of curved space in general relativity. It is invariant whatever the reference frame chosen. The measurement of time can be turned into a measurement of distance (t × c = ct); we can therefore say that time is space (or rather, motion through space). Nonetheless, there are big differences between time and space (John Wheeler). The central idea of relativity is that it is impossible to speak about quantities such as speed or acceleration without having chosen a reference frame. The geometrical description of physical theory given by Einstein has its origins in the advances of non-Euclidean geometry generalized by Bernhard Riemann as Riemannian geometry. Special relativity holds that this reference frame can be extended indefinitely in space and time. It only deals with so-called inertial reference frames. General relativity deals with reference frames whether accelerated or not, mathematically speaking. This equation underlies the famous expression that states that the curvature of space defines the motion of matter, and matter defines the curvature of space (the two being equivalent).

Transferred to physical space, the presence of a massive body will affect the curvature of space, which, seen from outside, will appear to alter the path of a light ray or of a moving object that travels close to it. General relativity is different from other existing theories due to the simplicity of coupling between matter and geometrical curvature.

As for time, it is always a local duration rather than universal time.

General relativity is a relativistic theory of gravitation. It refutes Newtonian gravitation and its concept of force. It states that gravitation is not a force but the expression of the curvature of space (in fact, of spacetime), a curvature which is itself produced by the distribution of matter. According to John Archibald Wheeler: "Matter and energy tell spacetime how to curve, and the curvature of spacetime tells matter how to behave." For Einstein, the motion of a body is not determined by forces but by the configuration of spacetime. This is a 'geometrization' of physics. If we take the example of the Earth and the Sun, according to general relativity it is the disturbance of spacetime caused by the matter of the Sun that results in the motion of the Earth.

A great number of experimental tests have been carried out, but none of them has succeeded in disproving general relativity, with the exceptions of the Pioneer anomaly, dark matter and the Casimir effect.

The equivalence principle
The Einstein equivalence principle which postulates the equivalence between gravitational mass and inertial mass can be stated as follows: "All reference systems in free fall are equivalent for the expression of non-gravitational physical laws, whatever their state of motion and their location." Reference frames in free fall can only be local. The Einstein equivalence principle, which is purely local, does not prevent the geometry of spacetime from changing from one point to another. On the contrary, such a change in geometry makes it possible to find a solution to the problem of gravitation. . In general relativity, the path of free particles such as photons is a straight line in a local inertial reference frame. Once these lines are extended beyond this local reference frame, they no longer appear straight, but are known as geodesics. The equivalence principle holds that there is no reason to locally distinguish free fall motion in a gravitational field from uniformly accelerated motion in the absence of a gravitational field. It is therefore natural to consider the motion of a particle in free fall in a gravitational field as being defined by a geodesic of a metric that is more complex than a Euclidean metric. In fact, Einstein introduced a so-called 'pseudo-Riemannian' generalization of special relativity's spatio-temporal metric. It models spacetime as a four-dimensional pseudo-Riemannian manifold and its gravitational field equation connects the curvature of the manifold at one point to the stress-energy tensor at that point, the tensor being a measure of the density of matter and energy (matter and energy being considered equivalent).

How can the curvature of spacetime created by a particular distribution of matter be calculated? It can be done thanks to Einstein's equations, which connect the curvature of spacetime to the distribution of matter. These equations are so complex that they can only be solved for very simple examples, such as a single isolated star.

The cosmological constant, Λ:

Einstein introduced the cosmological constant Λ so that a static Universe (neither expanding or contracting) would be one solution to his equations. Ten years later, when Edwin Hubble showed that the Universe was expanding, Einstein rejected the cosmological constant Λ from his equations, declaring that the introduction of the constant Λ was "the biggest blunder of his life".

The Einsteinian gravitational constant $8\mu G/c^2$

The absolute measurement of G was carried out by Cavendish (1731 – 1810) in 1798. The gravitational constant is $M^{-1} L^3 T^{-2} = 6.67259 \times 10^{-11}$ m^3 $kg^{-1} s^{-2}$.

Observational and experimental evidence for general relativity

1) The precession of the perihelion of Mercury.
2) The bending of light rays passing close to a mass was detected by observing stars near the Sun during an eclipse (1919).

3) Gravitational mirages observed (gravitational lensing by a large mass or by dark matter located on the path of light rays).

4) Gravitational clock drift taken into account in GPS satellites.

5) The prediction of gravitational waves: research is still under way, but some indirect evidence appears to support this prediction.

6) The concept of 'black holes' resulting from supernovae, which follows on from the equations of general relativity, is supported by indirect experimental evidence.

The concept of the cosmological constant, Λ, initially introduced by Einstein and then subsequently refuted by him, "the greatest blunder of his life", is highly questionable, as we have shown in several chapters, due to its cosmological observational inconsistencies.

Critique of the geometrical model of general relativity

The Einsteinian spacetime of general relativity, whose geometry is curved by the presence of mass-energy, is an anthropic concept. If it is a vacuum, then since by definition it has no properties, it cannot be curved. If it is not a vacuum, is it a mathematical space or a physical space? If this space is mathematical, what is its connection with physical reality? If it is physical, and can be curved by mass-energy, what differentiates it from a vacuum?

The temporalistic model: a physical model

Unlike the Newtonian parameter G or the Einsteinian gravitational constant, $8\mu G/c^2$, temporalistic gravitation interprets gravity as being governed by the temporalistic parameter G'. The existence of the temporalistic field implies that of an energy field in space, even in the absence of particles of matter or of energy.

How can we determine the energy state corresponding to the existence in space of the temporalistic field? This is where we see the deep significance and justification of the dimension of the parameter G' that we used earlier. We formulated the hypothesis of a dimension LT^{-2} (that of an acceleration) for the gravitational constant G' in the temporalistic model, whereas the Newtonian dimension is $M^{-1}L^3T^{-2}$. In the Newtonian and relativistic theories, the gravitational constant G (or $8\mu G / c^2$) is a parameter related to mass and energy. This parameter, as we recalled earlier, gives the strength of the coupling between masses, or between masses (and energy) and the metric field (spacetime).

In classical theories of gravitation, gravitational interaction has an infinite range. In the temporalistic model, the range of disturbance of the gravity field of gravitons by masses is limited by the value of the universal acceleration field, that is, G' (6.582×10^{-10} m/s² in SI or 6.582×10^{-8} cm/s² in the CGS system). The disturbance caused to the universal acceleration field by the presence of masses and energy becomes apparent via the emergence of a local acceleration field. This screening or disturbing effect will only be felt if it is greater than the universal acceleration field. In other words, if the strength of the disturbance caused to the universal acceleration field by the screening effect of the masses is less than G', the disturbing effect of the masses will no longer be felt. Temporalistic gravitation therefore has a limited range. This upper spatial limit is given by the approximate formula $r = m^{1/2}$. This is the gravitational radius of masses.

In fact, the Einsteinian concept of spacetime can be interpreted ananthropically, i.e. without being irrational or inconsistent, using the model of gravitation proposed by the author: <http://site.voila.fr/nobigbang> In this model, the vacuum is filled with a universal acceleration field of gravitons; matter and energy (stars, galaxies, clouds of dust and gas, galaxy clusters and superclusters, etc) are the disturbance factors of this universal acceleration field of gravitons. They locally distort this (physical) acceleration field. This distortion corresponds to the (geometrical) curvature of spacetime in general relativity.

The temporalistic model is not in conflict with general relativity. It interprets the Einsteinian (mathematical) concept of a curved spacetime as a local (physical) distortion, caused by mass and energy, of the universal acceleration field of gravitons. The temporalistic model adds to it a fundamental element: the range of gravitation is finite.

Validation of this model has been provided by a large number of verifications (Chapter XII: Masses and gravitational radii).

Masses and gravitational radii

Observational evidence for temporalistic gravitation

We have seen that the range of the gravitational radius of masses postulated by the temporalistic model is given by the expression $r = m^{1/2}$ (r = radius, m = mass). The gravitational radius marks the outer limit of gravitational interaction between masses, whether small or great (such as stars, galaxies, clusters, etc), i.e. the range of the gravitational radius of a mass may only be equal to or less than r, in other words than the acceleration G' (6.582×10^{-8} cm/s^2) of the gravity field. In Newtonian mechanics, the acceleration caused by masses is also mG / L2. Thus, if we apply this expression to the Milky Way (mass approximately 2 to 3×10^{45} g, radius 50 000 ly, we get (in the CGS system): 2.5×10^{45} g $\times 6.67 \times 10^{-8}$ cm/s^2 / $5 \times 10^{22} \times 5 \times 10^{22}$ cm = 6.67×10^{-8} cm/s^2. At the outer edge of our Galaxy, acceleration is equal to the acceleration of the external gravity field. It is therefore neutralized, and according to the temporalistic model, its gravitational radius is indeed given by the expression $r = m^{1/2}$ (Chapter XII), that is $(2.5 \times 10^{45})^{1/2} = 5 \times 10^{22}$ cm.

We shall calculate, for known concentrations of mass, from the planet Earth to the largest structures in the Universe, the theoretical gravitational radius, of finite range, and compare it with the observed sizes of these different masses (in the CGS system). When the masses are not accurately known, we have estimated the total mass of a structure as being approximately equal to around ten times the visible mass (in agreement with estimations that there is 4% visible matter and 24% dark matter: 28 / 4 = 7, that is around 10 times).

Before getting to the heart of the matter, we need to remember that the masses and distances of these different structures, especially when they are distant, are only known very approximately. For instance, Zwicky (1933) pointed out that the ratio of the dynamic mass of a structure to its visible mass was in the region of 400. In addition, it is generally admitted that the constant of proportionality between redshift and distance is generally only known to the nearest factor of 2.

Taking all these uncertainties into account, we shall compare the theoretical gravitational radius of various known cosmic masses with their real gravitational radius, determined either from the sizes of these masses or by the limit of their effect on other masses. We shall consider that if we occasionally obtain a difference in order of magnitude, then this difference is acceptable given the approximate nature of cosmic parameters. For masses that are not precisely known, the estimated mean masses were multiplied by 10, due to the presence of dark matter.

1) The Earth: mass 6×10^{27} g – gravitational radius 7.7×10^{13} cm – distance of the Moon 3.5×10^{10} cm – magnetosphere – approximately 8×10^9 cm (Philippe Escoubet 2001).
2) The Solar System: mass of the Sun 2×10^{33} g – gravitational radius 4.5×10^{16} cm – boundary of the Solar System with interstellar space 1.4 to 1.8×10^{15} cm (NASA 1993), heliopause 4.5×10^{15} cm – Oort Cloud, influenced by the stars of the Milky Way, 3×10^{18} cm (Rosanna L. Hamilton 1999). The Kuiper belt is currently at a distance of around 34 AU, or 5×10^{14} cm; the interface between the solar wind and interstellar gas which extends for several hundred million km is the heliopause (around 94 AU, or 9.4×10^{14} cm). The termination shock between the solar wind and ionized interstellar gas is located at a distance in the region of 100 AU (1.5×10^{15} cm) (Dossier Pour la Science N° 64 – July-September 2009).
All these measurements are consistent.
3) Globular clusters:
Average mass of 10 000 stars, that is, 2×10^{33} g $\times$ 10 000 = 2×10^{37} g – total estimated mass 2×10^{38} g – gravitational radius 1.4×10^{19} cm – average radius several tens of AU, or 2 to 3×10^{19} cm (Hartmut Frommert - Christine Kronberg - 2001).
Average mass of 1 million stars, that is, 2×10^{33} g $\times 10^6$ = 2×10^{39} g – total estimated mass 2×10^{40} g – gravitational radius 1.4×10^{20} cm – average radius 200 ly, or 2×10^{20} cm (Hartmut Frommert - Christine Kronberg - 2001).
M92 – estimated mass of around 330 000 stars, that is, 2×10^{33} g $\times$ 330 000 = 6.6×10^{38} g – gravitational radius 2.6×10^{19} cm – radius 30 to 42 ly, that is, 3 to 4×10^{19} cm (Hartmut Frommert - Christine Kronberg - 2001).

4) The Milky Way: 200 billion stars, that is, $2 \times 10^{11} \times 2 \times 10^{33}$ g = 4×10^{44} g, estimated mass 4×10^{44} g $\times$ 10 = 4×10^{45} g – gravitational radius 6.3×10^{22} cm - radius 50 000 ly, that is, 5×10^{22} cm – dwarf satellite galaxy SagDEG at 5×10^{22} cm (Hartmut Frommert - Christine Kronberg - 1999); the satellite galaxies of the Milky Way, the Small and Large Magellanic Clouds are located 60 kpc from our Galaxy, that is, 2×10^{23} cm.

5) Galaxy clusters: Typical cluster 10^{15} solar masses, that is, 2×10^{33} g $\times 10^{15}$ = 2×10^{48} g – gravitational radius 1.4×10^{24} cm – typical Abell radius 1.5 Mpc, that is, 5×10^{24} cm (Coma cluster) (Cambridge Cosmology).

We have used the following figures, taken from a consensus by specialists:

Group of 10 galaxies: average mass 10^{13} solar masses, that is, 2×10^{46} g (average mass of a galaxy 2×10^{45} g).
Standard cluster: 500 galaxies, average mass 3×10^{14} solar masses, that is, 6×10^{47} g (average mass of a galaxy 1.2×10^{45} g).
Rich cluster: 3 000 galaxies, average mass 5×10^{15} solar masses, that is, 1×10^{49} g (average mass of a galaxy 3×10^{45} g).
We have therefore decided to accept a mass of 2×10^{45} g as the average mass of a galaxy with an average gravitational radius = 5×10^{22} cm.

6) Virgo cluster: estimated mass 8×10^{48} g – gravitational radius 3×10^{24} cm – maximum distance of galaxies from the center of the cluster: 7 million ly, that is, 7×10^{24} cm.

7) Galaxy superclusters: 10 to 32 clusters per supercluster on average – Our supercluster (which contains the Local Group), is centered on Virgo, mass 10^{16} solar masses, that is, 2×10^{33} g $\times 10^{16}$ = 2×10^{49} g – the matter/luminosity ratio is 570, which points to the presence of a large amount of black matter – probable gravitational radius 4.5×10^{24} cm / 1×10^{25} cm (around 1.5 to 3 Mpc) – radius 2×10^{25} cm (Cambridge Cosmology).

8) The Great Attractor: a super supercluster whose center is the supercluster ACO 3627 (or Norma cluster), mass 5×10^{16} solar masses, that is, 2×10^{33} g $\times 5 \times 10^{16}$ = 1×10^{50} g (its mass is probably greater; the existence of other undetected superclusters is suspected) – gravitational radius 1×10^{25} cm – distance from the Earth 60 Mpc, that is, 1.8×10^{26} cm. The data are uncertain, due to the fact that the Great Attractor was long hidden by the dust of the Milky Way disk. (Kraan-Korteweg 1998 - 2000).

9) The Large-scale Structures of the Universe: The galaxies, made up of stars, gas, dust and dark matter, are grouped together in galaxy clusters, then in galaxy superclusters, themselves arranged into huge formations such as great walls, filaments and great voids. Depending on the authors, the Universe has a structure similar to foam, sponge, sheets, pancakes or a

three-dimensional spider's web. In fact, it can be considered that the Universe is organized into filaments made up of gas, dust, stars, galaxy clusters and superclusters, dark matter and great voids. The standard model of cosmology is unable to explain the reasons for the existence of these structures and huge voids. On the other hand, the temporalistic model offers a simple explanation for the structure of the Universe and for the existence of filaments and great voids. Chapter IX (The evolution of galaxies – The large-scale structures of the Universe) sets out this proposition in more detail.

10) Average gravitational radii and average distances:
Stars in galaxies: gravitational radius 4×10^{16} cm – average distance 1 pc, that is, 3×10^{18} cm.
Galaxies in groups and clusters: gravitational radius 4×10^{22} cm – average distance 1 Mpc, that is, 3×10^{24} cm.
Galaxy clusters in superclusters: gravitational radius 1.4×10^{24} cm – average distance 1 to 10 Mpc, that is, 3×10^{24} cm to 3×10^{25} cm.
Galaxy superclusters: gravitational radius 5×10^{24} cm to 1×10^{25} cm – average distance 100 Mpc, that is, 3×10^{26} cm.
Voids have average sizes of 1×10^{26} cm or larger (1×10^{27} cm).

Conclusions: If we summarize these results, we can see that, in agreement with the requirements of the temporalistic model, the size and the gravitational effect of concentrations of matter, from the Earth right up to the largest structures, are, in order of magnitude, equal to or less than the gravitational radii. The only exception is the Great Attractor, to the nearest order of magnitude. It is likely that its mass or its distance, or both, will need to be revised. This is particularly likely since the Great Attractor is hidden by the dust in the Milky Way disk, which alters the accuracy of measurements. The size of voids, in the region of 10^{26} cm and more, can also be explained by the smaller gravitational radius of galaxy superclusters, in the region of 1×10^{25} cm.

Classical theories of gravitation in which the range of forces is unlimited, as well as the Big Bang theory, can explain neither these results nor their accuracy. The Universe appears to be structured with a periodicity of distribution in three dimensions, separated by voids with average sizes of 120 Mpc (4×10^{26} cm), as on a chessboard. These structures, which are inexplicable in other models, result naturally from the finite range of gravitational radii which is specific to the temporalistic model of gravitation.

Moreover, the formation of these great voids poses a serious problem to the Big Bang model. To cross a void with a size in the region of 4×10^{26} cm, at the average speed for a galaxy of 600 km/s, would take about 200 billion years, which means that the current location of galaxies and voids reflects their location at the time of the Big Bang!

Chapter XII validates, for masses ranging from that of the Earth to those of the largest structures in the Universe (galaxy superclusters, great voids, etc), the relationship between their mass and their gravitational radius.

The temporalistic model proposes a Universe with neither a beginning nor an end, which has many consequences. It makes it possible to solve a large number of problems that are specific to the Big Bang model. It proposes a series of precise tests capable of confirming the model or invalidating it (Chapter XIV).

We shall see, in the conclusions, that Olbers' paradox as well as many other problems in the Big Bang model, find a natural solution from the perspective of the temporalistic Universe. Cosmologically speaking, this is shown to be a Universe that is spatially relatively static, but temporally dynamic and evolving.
The value of the temporalistic effect or 'recession effect' at 1 Mpc = 67.71 km/s and that of Ho = 4.5546×10^{17} s (approximately 14.43 billion years) were <u>established theoretically by the author in 1962.</u>

The latest data provided by WMAP 5 (Table 7 – Cosmological Parameter Summary – 2008) gives a value for Ho = 71.9 (+2.6 – 2.7) km/s/Mpc and to = 13.69 (± 0.13) billion years.
Let us recall the values of Ho and To that we obtained in Chapter VII. Comparing the observational value and the theoretical value for Ho: 69.2 km/s/Mpc (71.9 – 2.7) for the former and 67.71 km/s/Mpc for the latter, there is a difference of 2.16%. This difference is negligible if we consider the uncertainty in the WMAP 5 data: between 3.2% (+2.6) and 3.75% (-2.7). We should add that the value of Ho provided by WMAP 5 was obtained after 80 years of research and corrections, of which 69.2 km/s/Mpc is the most recent but certainly not the final result, whereas the theoretical value proposed by the author as long ago as 1962 has not changed since then. The value for the Hubble Constant, Ho, provided by NASA (2008) is the result of a great many cosmological observations and the unremitting work of a vast number of researchers. The SDSS (Sloan Digital Sky Survey) project, which has studied the redshift of 221 414 galaxies, has not modified this estimation. However, due to the very nature of the observations, the accuracy of the results can only be relative (like, for

instance, the distance of distant celestial bodies such as galaxies or galaxy clusters, whereas the value of the Ho constant theoretically established and proposed by the author is very accurate since it is based on the values of the universal and/or quantum constants that he uses as well as on their accuracy (c, G, h, e).

From Hubble's Law, $v = Ho \times d$, where v = recession speed in km/s, Ho = Hubble's constant in km/s/Mpc and d = distance in Mpc, we get Ho = v / d = 69.2 km/s / 3.084 × 10^{19} km (3.15576 × 10^7 s × 10^6 × 3.26 × 2.997925 × 10^5 km/s) = 2.243 × 10^{-18} s. If the Universe has a very low matter density, which is the case, the age of the Universe, to, equals 1 / Ho = 1 / 2.243 × 10^{-18} s = 4.458 × 10^{17} s, which is around 14.12 billion years. The differences with the values obtained by the author are, as for the values of Ho, in the region of 2.15% (Ho = 67.71 km/s/Mpc and To = 4.5546 × 10^{17} s), in other words within the range of the uncertainties.

Latest news: A new planet, the most distant one discovered, located beyond Pluto at a distance of 97 astronomical units, that is, 1.45 × 10^{15} cm, therefore within the gravitational radius of the Sun, 4 × 10^{16} cm, has just been discovered (Pour La Science N° 410 December 2011 – page 12).

Chapter X: Dark matter –The Pioneer effect – The MOND theory – The Casimir effect confirm, in these areas, the temporalistic model qualitatively and/or quantitatively.

PART SIX

General conclusions

Chapter XIII

Summary of critiques concerning the various concepts used by the standard Big Bang model

The Big Bang delusion

Redshifts

(Chapter VIII)

The expansion of the Universe and the recession of galaxies are not observational data. They result from an interpretation of the redshift of distant galaxies, which are interpreted, in the standard Big Bang model, as being a spatial cosmological effect caused by the expansion of the Universe. The concept of the expansion of physical space amounts to assigning properties to the vacuum which, by definition, it cannot have, such as curvature. The same applies to the concept of time, which disappears, without any explanation, at the speed of light.

The temporalistic model, based on the quantum constant To, proposes an alternative to the Big Bang model and a new interpretation of redshifts. The temporalistic model interprets redshifts as a <u>quantum and temporal</u> phenomenon, and not as a <u>cosmological and spatial</u> one. According to the temporalistic model, the redshift, z, of photons traveling through space, is the result (aside from any external interaction) of the influence of the asymmetry of time and of the existence of the temporalistic constant To, on the parameters of photons. It has no relationship with the concept of 'tired light'.

Let us recall the latest data provided by WMAP 5 (Table 7 – Cosmological Parameter Summary – 2008) which give a value for Ho = 71.9 (+2.6 – 2.7) km/s/Mpc and to = 13.69 (± 0.13) billion years.

Comparing the observational value and the theoretical value of Ho: 69.2 km/s/Mpc (71.9 – 2.7) for the former and 67.71 km/s/Mpc for the latter, there is a difference of 2.16%. This difference is negligible if we consider the uncertainty in the WMAP 5 data: between 3.2% (+2.6) and 3.75% (-2.7). We should add that the value of Ho provided by WMAP 5 was obtained after 80 years of research and corrections, of which 69.2 km/s/Mpc is the most recent but certainly not the final result, whereas the theoretical value proposed by the author as long ago as 1962 has not changed since then. Its accuracy is based on that of the universal quantum constants he uses (c, G, h, e).

<u>This theoretical value was obtained using purely theoretical methods and is independent of any astronomical data, which strengthens its validity.</u>

<u>The Cosmic Microwave Background (CMB)</u>

(Chapter VIII)

The cosmic microwave background is not <u>evidence</u> for the existence of the Big Bang. Once again, it is merely an <u>interpretation</u> of a factual phenomenon, in correlation with a hypothetical model, the Big Bang model. The cosmic microwave background is interpreted as <u>fossil radiation</u> dating back to 380 000 years after the Big Bang Once again, this interpretation, can be seen to be a simple <u>hypothesis</u> rather than <u>evidence</u> for the Big Bang model.

The hypothesis of the cosmic microwave background, used, wrongly, as a validation of the hypothesis of the existence of the Big Bang raises a number of problems:

the horizon problem
the problem of the flatness of the Universe (almost the same as its critical density)
the problem of the homogeneity and isotropy of the Universe
the singularity problem

The various problems raised by the interpretation of the cosmic microwave background as being so-called evidence for the phenomenon of the Big Bang have forced the supporters of the Big Bang model to formulate a new *ad hoc* hypothesis, the concept of inflation, which is far more questionable and hypothetical, has no experimental or observational foundation, and violates the laws of current physics. A major argument that leads to the rejection of inflationary theories is their ability to fit all possible initial conditions (See the highly negative opinion of James Peebles, who supports the Big Bang model 'by default', about inflation: 'Inflationary theories', page 83.

Primordial nucleosynthesis

(Chapter VIII)

The agreement between the predictions of abundances of light nuclei using the basic hypotheses of the Big Bang and the current abundances of these nuclei is supposed to be one of the strong points of the standard Big Bang Nucleosynthesis model . Conversely, their lack of agreement calls into question the standard Big Bang Nucleosynthesis model.

All the hydrogen and part of the helium and lithium contained in the Universe are supposed to have been formed in the first hundred seconds after the Big Bang. According to the theory, a comparison of the results from the latest theoretical calculations concerning nucleosynthesis and from WMAP 5 data show that the values inferred from the cosmic microwave background and astrophysical observations <u>agree</u> for deuterium, are <u>no more than reasonable</u> for helium-4, but are <u>in complete disagreement for lithium-7</u>.

These serious discrepancies call into question the standard model of Big Bang nucleosynthesis.

According to the supporters of the standard Big Bang model, redshift, the cosmic microwave background and primordial nucleosynthesis make up the three pillars of the theory.

<u>We can only note that these so-called pillars have great difficulty standing up to the critical analysis that we have carried out and whose results, which are indisputable, fly against the Big Bang model and destroy its credibility.</u>

The horizon problem

(Chapter VIII)

In order to solve the horizon problem, the standard Big Bang model of cosmology requires a new hypothesis, the highly speculative hypothesis of inflation, a non 'falsifiable' hypothesis whose many difficulties were analyzed and which, far from solving the horizon problem, simply piles up fresh problems.

Solving the horizon problem by postulating the existence in the Universe of a form of matter that has negative pressure, for which there is no experimental or observational validation, merely constitutes another *ad hoc* hypothesis without any empirical or theoretical justification.

The hypothesis of the existence in the Universe of a form of matter that has negative pressure brings us back to the cosmological constant Λ, considered similar to vacuum energy, which is known to result from the predictions of quantum field theory, but which leads to a totally unacceptable value, 60 to 120 times greater than the value inferred from cosmological observations.

The problem of flatness and critical density

(Chapter VIII)

Inflationary theories, which are highly speculative, are supposed to solve the flatness problem even though they are themselves a source of serious difficulties. The solution proposed for the flatness problem, which is identical to the solution to the horizon problem, namely inflation, therefore suffers from the same difficulties. In other words, it is a highly speculative *ad hoc* hypothesis that relies on the concept of matter with negative pressure, which lacks any observational backing, and with a totally unacceptable value, 60 to 120 times greater than the value inferred from cosmological observations

The problem of the homogeneous, isotropic Universe

(Chapter VIII)

Inflationary theories are supposed to solve the problem of the homogeneous and isotropic Universe, as well as the horizon and flatness problems, even though they are themselves a source of serious difficulties Inflation therefore suffers from the same difficulties, in other words it is a highly speculative *ad hoc* hypothesis, with the concept of matter with negative pressure, which lacks any observational backing, and with a totally unacceptable value, 60 to 120 times greater than the value inferred from cosmological observations

According to James Peebles, an eminent supporter of the Big Bang model as we saw earlier, a major argument that leads to the rejection of inflationary theories is their ability to fit all possible initial conditions (Chapter VIII: inflationary theories, page 86).

The singularity problem and the origin of the Big Bang

(Chapter VIII)

According to the Big Bang model, the Universe was born, in the 'primordial explosion', from a spacetime singularity which had an 'infinite' density and temperature. What was the cause of this explosion? No answer to this question is provided by the current laws of physics. Or else, this difficulty is side-stepped by denying the 'primordial explosion', without any clear or valid justification. Where did space, time, matter and energy come from? Apparently, they were created *ex nihilo*. Yet again, this is merely a simple assertion, without any experimental or factual validation.

Until now, no *ex nihilo* creation of matter or energy has ever been observed, whether in physical or biological phenomena. To assert that space and time appeared with the Big Bang is a circular argument which arbitrarily eliminates, without any evidence, the problem of the existence of time before the Big Bang. Some even assert, against all logic, that the Big Bang happened "nowhere and everywhere at the same time".

It is also equally possible to assert, without any other justification, that at the singularity of the Big Bang, the notion of space disappears but not that of time (Gabriele Veneziano's pre-Big Bang). As usual, this new hypothesis is not 'falsifiable'.

Many other hypotheses have been put forward: All these models are highly speculative, without there being any possibility of validating them. This in no way bothers their authors, who claim the right to speculate without the need for any restrictive tests.

When all is said and done, the Big Bang model is a strictly anthropic concept. It is irrational and speculative, at the expense of critical thinking. It infringes current physical laws without providing any experimental or observational validation.

Inflationary theories

(Chapter VIII)

Inflationary theories are an extension of the Big Bang model, but are independent of it.

The inflationary model, created in order to solve the problems of the Big Bang model, (horizon problem, flatness problem, homogeneity problem, etc) is, in the final analysis, nothing more than an *ad hoc* hypothesis without any experimental or factual basis. The considerable extrapolation of the laws of physics set out in this model has no theoretical justification, apart from providing an arbitrary response to the difficulties of the Big Bang model. A major argument that leads to the rejection of inflationary theories is their ability to fit all possible initial conditions.

We can do no more than repeat the harsh judgement of the eminent researcher James Peebles, a supporter of the Big Bang:

"The assertions of the inflationary model, which are poorly demonstrated, can lead to genuine scepticism in the eyes of rigorous observers." (James Peebles 2001). "It is a theory that can be adjusted in order to produce the structures that we see, starting out from all the possible initial conditions. From this perspective, it is not really a theory but an 'off the peg' story, since it fits all cases. All you have to do is change a few parameters." This *ad hoc* mechanism only has value by default. "In any case, we don't have a better one." (James Peebles - Les Dossiers de la Recherche – N° 35 – (Quarterly) May 2009 – page 8).

The acceleration of expansion – Dark energy

(Chapter VIII)

The acceleration of expansion leads to the hypothesis of the existence of dark energy. Different models propose an explanation for this: a) the cosmological constant Λ. However, the predictions of quantum field theory lead to a totally unacceptable value, 60 to 120 times greater than the value inferred from cosmological observations; b) quintessence, which was abandoned several years ago due to its numerous problems; c) general relativity, with its 'scalar tensors': no observations have been able to validate this concept, which remains purely hypothetical; d) axions: this model has now been abandoned. None of the models proposed is either 'falsifiable' or validated. In desperation, some have had no hesitation in proposing an 'anthropic model'!

To sum up, the concepts of acceleration of expansion and dark energy lead to so many problems, without any valid solutions, that it appears incongruous to use them. The Big Bang theory, which incorporated them into its standard model, therefore suffers from the same unavoidable difficulties as do these concepts.

Only inhomogeneous and non-isotropic models of the Universe, with their questioning of the cosmological principle, are free from this critique. They lead to a rejection of accelerating expansion and its consequence, the existence of dark energy.

The theoretical prediction of the Hubble constant, Ho

The age of the Universe, to

(Chapter VII – Chapter V)

Estimating the age of the Universe by studying its components (stars, globular clusters, galaxies, atomic nuclei, etc) gives orders of magnitude that are far too great, ranging from 11 to 18 billion years.

The latest data provided by WMAP 5 (Table 7 – Cosmological Parameter Summary – 2008) gives a value for Ho = 71.9 (+2.6 – 2.7) km/s/Mpc and to = 13.69 (± 0.13) billion years.

When we compared the observational value and the theoretical value for Ho: 69.2 km/s/Mpc (71.9 – 2.7) for the former and 67.71 km/s/Mpc for the latter, we obtained a difference of 2.16%. This difference is negligible if we consider the uncertainty in the WMAP 5 data: between 3.2% (+2.6) and 3.75% (-2.7). the value of the Ho constant theoretically established and proposed by the author is very accurate since it is based on the accuracy of the universal and quantum constants that he uses (c, G, h, e).

In conclusion, the temporalistic model refutes the interpretation of the origin of redshifts as being the expansion of space, and interprets redshifts (the increase in wavelength of moving photons) as being physical phenomena caused by the existence of the temporalistic constant, To, with a value of 4.5546×10^{17} s.

(Chapter IX)

The evolution of galaxies – The large-scale structures of the Universe

The model of creation and evolution of galaxies and large-scale structures in the Big Bang model raises a very large number of problems: what happened before the Planck time (10^{-43} seconds)? What was the process of creation of matter? From nothing? How? What was the cause of the Big Bang? The redshift of distant galaxies, revealed by Hubble, on which the standard model of cosmology is based, implies a singularity with temperature, density and energy parameters with exceptionally high values. This singularity cannot be incorporated into today's physics, since the equations of both general relativity and of quantum field theory can no longer be used due to the appearance of many infinite terms (see the Origin of the Big Bang - Chapter VIII).

The energy fluctuations that arose several thousand years after the Big Bang, from which galaxies are supposed to have formed through the action of gravity, are not enough to justify the evolution of large-scale structures. According to Tegmark (2004), although the anisotropies in the cosmic microwave background are entirely in accordance with this idea on small and medium scales, this is not at all true on large scales. The way in which the structures develop depends on the origin of the primordial fluctuations and on the nature of dark matter.

Depending on the authors, the Universe has a structure similar to foam, sponge, sheets, pancakes or a three-dimensional spider's web. To sum up, it may be considered that the large-scale structures of the Universe are made up of filaments made of gas, dust, stars, galaxies, clusters and superclusters of galaxies, great walls, great voids and dark matter. Great voids, the probability of whose existence is estimated to be 5×10^{-10}, as well as the various other inhomogeneous structures already discovered, seriously challenge the standard model of cosmology, based on <u>the cosmological principle, which gives the Universe a homogeneous and isotropic structure</u>.

Summary of the different critiques

The Big Bang model results, in all its different aspects, from the spatial interpretation of the origin of redshifts, i.e. the spatial expansion of the Universe. You could say that this is its '<u>original sin</u>', which has led researchers into a genuine dead end.

In fact, redshifts have an origin that is related to <u>time</u> and not to <u>space</u>.

The interpretation of redshifts is of fundamental importance, since it makes up the premises on which the standard Big bang model is based. If the temporalistic interpretation turns out to be correct, then all the interpretations and all the concepts of the Big Bang model collapse.

Redshifts result from the nature of photons, which are affected, as they travel through space, by the existence of the 'temporalistic constant' To whose value is 4.5546×10^{17} s.
All the concepts and all the difficulties of the Big Bang model that we have analyzed are caused by the paradigm of the expansion of space, which leads to highly speculative hypotheses (inflationary theories, the multiverse, singularities, etc), which violate the laws of physics and of logic (*ex nihilo* creation of mass-energy, primordial explosion at the origin of spacetime, curvature of space, i.e. of nothingness, which is an oxymoron, etc). The Big Bang model has led to an ever growing number of speculations by cosmologists, together with a clear rejection of the usual rigorous rules of science, Popper's 'falsifiability' and Einstein's 'observable facts'.

The temporalistic model is in complete conflict with the concepts and methodologies of the Big Bang model. It results from a single hypothesis, the existence of the 'temporalistic constant', To, from which it draws all its conclusions, strictly respecting the requirements of Popper's 'falsifiability' and Einstein's 'observable facts'. Because it requires a rigorous approach, the temporalistic model sidesteps all the difficulties of the Big Bang model (such as singularities, *ex nihilo* creation of mass-energy, etc).

Contrary to the assertions of the supporters of the Big Bang model, the cosmic microwave background is not evidence for the existence of the Big Bang. Once again, it is merely an interpretation of a factual phenomenon, in correlation with a hypothetical model, the Big Bang model. The cosmic microwave background is interpreted as fossil radiation dating back to 380 000 years after the Big Bang. Once again, this interpretation, can be seen to be a simple hypothesis rather than evidence for the Big Bang model.

A comparison of the results from the latest theoretical calculations of nucleosynthesis and from WMAP 5 data show that the values inferred from the cosmic microwave background and astrophysical observations agree for deuterium, are no more than reasonable for helium-4, but are in complete disagreement for lithium-7. All these discrepancies call into question the standard model of Big Bang nucleosynthesis.

Redshift, the cosmic microwave background and primordial nucleosynthesis, which make up, according to the supporters of the standard Big Bang model, the three pillars of the model, emerge much weakened from the critical analysis we have subjected them to. We have seen that mere hypotheses are interpreted as, and stated to be, evidence (redshifts, the cosmic microwave background), while other observations show serious discrepancies (primordial nucleosynthesis). These major problems, hidden by the Big Bang model, destroy its credibility.

The hypothesis of the reality of the Big Bang leads to a certain number of additional difficulties:
The problems of the horizon, of the flatness of the Universe (almost the same as its critical density) and of the homogeneity and isotropy of the Universe undermine the Big Bang model.

Inflationary theories are supposed to solve these three problems, even though they are themselves sources of serious difficulties. Inflation therefore suffers from the same difficulties, in other words it is a highly speculative hypothesis, with the concept of a cosmological constant Λ and of matter with negative pressure, lacking any observational backing, and with a totally unacceptable value, 60 to 120 times greater than the value inferred from cosmological observations It provides the problems arising from the Big Bang model with a solution that is merely an *ad hoc* hypothesis, in other words, it replaces these problems with other, even more insurmountable problems.

To get round the various difficulties of the Big Bang model, cosmologists have transferred them to another, even more hypothetical and questionable concept, inflation. In fact, this is to jump out of the frying pan into the fire.

"It (inflation) is a theory that can be adjusted...since it fits all cases. All you need to do is to change a few parameters." (James Peebles - Dossier trimestriel N° 35- May 2009 - La Recherche – page 8).

The singularity problem and the origin of the Big Bang: all the solutions proposed to the problems of the singularity and the origin of the Big Bang are irrational and speculative; they infringe the current laws of physics, at the expense of critical thinking, without providing any credible experimental or observational validation. They are strictly anthropic concepts.

The acceleration of expansion – Dark energy: the concepts of accelerating expansion and dark energy lead to so many problems, without any valid solutions, that it appears incongruous to use them. The Big Bang model, which incorporated them into its standard model, therefore suffers from the unavoidable problems caused by these concepts.

The theoretical prediction of the Hubble constant, Ho – the age of the Universe, to: the latest data provided by WMAP 5 (Table 7 – Cosmological Parameter Summary – 2008) gives a value for Ho = 71.9 (+2.6 – 2.7) km/s/Mpc and to = 13.69 (± 0.13) billion years.

Comparing the observational value and the theoretical value for Ho: 69.2 km/s/Mpc (71.9 – 2.7) for the former and 67.71 km/s/Mpc for the latter, gives a difference of 2.16%. This difference is negligible if we consider the uncertainty in the WMAP 5 data: between 3.2% (+2.6) and 3.75% (-2.7). We should add that the value of Ho provided by WMAP 5 was obtained after 80 years of research and corrections, of which 69.2 km/s/Mpc is the most recent but certainly not the final result, whereas the theoretical value proposed by the author as long ago as 1962 has not changed since then. Its accuracy is based on that of the universal quantum constants that he uses (c, G, h, e). (See Chapter VII – Chapter VIII).

The evolution of galaxies – The large-scale structures of the Universe: The energy fluctuations that arose several thousand years after the Big Bang, from which galaxies are supposed to have formed through the action of gravity, are not enough to explain the evolution of large-scale structures. According to Tegmark (2004), although the anisotropies in the cosmic microwave background are entirely in accordance with this idea on small and medium scales, this is not true at all on large scales. The way in which the structures develop depends on the origin of the primordial fluctuations and on the nature of dark matter.

In Chapter XIV, the author compares the two conflicting models, the standard Big Bang model and the temporalistic model, together with their strengths and weaknesses. It is up to the reader to make up his/her own mind as to which model appears to be the most pertinent and most scientifically valid.

THE BIG BANG DELUSION

We have just summarized the weaknesses and/or inconsistencies of all the concepts in the Big Bang model, even though this theory currently enjoys near-dogma status.

As we have pointed out several times, historically, the Big Bang model is the consequence of Edwin Hubble's interpretation in 1929, of the redshifts of distant galaxies as being the 'recession of galaxies'. This interpretation led, quite naturally, to the concept of the expansion of space, or rather of spacetime, i.e. of the Universe. On the basis of these premises, the Big Bang model grew for 80 years, with the creation and development of many concepts that were supposed to validate it, starting out from a large number of observations and hypotheses: the cosmic microwave background, primordial nucleosynthesis, dark matter, dark energy, etc. Because of the erroneous nature (in our opinion, the 'original sin') of the premises of the theory, the spatial interpretation of the redshifts of distant galaxies, it was inevitable that problems would arise. Indeed, problems did not fail to arise, due to discrepancies and questions raised by observations (the 'three pillars' of the Big Bang, the problems of the horizon, of the flatness of the Universe – almost equal to its critical density – , of the homogeneity and isotropy of the Universe, etc).

The new *ad hoc* inflationary hypotheses, aimed at solving the previous problems, were only able to solve them at the expense of new highly speculative hypotheses that were compromised by violations, for which there were no serious justifications, of current physical laws (exponential speed of the expansion of the Universe, well over the upper speed limit c (special relativity) allowed by contemporary physics, unexplained causes and modes of inflation, etc) and whose credibility has been widely questioned by an eminent cosmologist, a supporter of the Big Bang model by default, James Peebles (Dossier trimestriel N° 35- Mai 2009 - La Recherche – page 8).

The problems of the singularity and of the origin of the Big Bang are not solved by the standard model of cosmology. This is because, since with general relativity and quantum field theory it is not possible to go back beyond the Planck time (10^{-43} seconds), the solution to this problem is quite simply hidden from us.

The concept of the acceleration of expansion and of dark energy (which results from the observation of type Ia supernovae) is controversial since it assumes a homogeneous and isotropic Universe that is itself questioned. Many models that are supposed to justify this concept have been proposed. Until now, only the model of the cosmological constant Λ has been accepted.

Unfortunately, the predictions of quantum field theory for the value of the cosmological constant give a totally unacceptable value, 60 to 120 times greater than the value inferred from cosmological observations. Other specific difficulties undermine the concepts of accelerating expansion and of dark energy (there is nothing standard about the formation of type Ia supernovae – for instance, supernova SN2006gz – and this may distort measurements made by cosmologists; according to observations of nearby supernovae (less than a billion light years away), the acceleration of expansion may have been decreasing over the past 2.5 billion years, to the point of reversing recently). The Big Bang model, which incorporated the concepts of accelerating expansion and dark energy into its standard model, is naturally affected by all these difficulties.

The theoretical prediction of the Hubble constant, Ho – the age of the Universe, to:

Let us recall the latest data provided by WMAP 5 (Table 7 – Cosmological Parameter Summary – 2008) which give a value for Ho = 71.9 (+2.6 – 2.7) km/s/Mpc and to = 13.69 (± 0.13) billion years.

Comparing the observational value and the theoretical value for Ho: 69.2 km/s/Mpc (71.9 – 2.7) for the former and 67.71 km/s/Mpc for the latter, there is a difference of 2.16%. This difference is negligible if we consider the uncertainty in the WMAP 5 data: between 3.2% (+2.6) and 3.75% (-2.7). We should add that the value of Ho provided by WMAP 5 was obtained after 80 years of research and corrections, of which 69.2 km/s/Mpc is the most recent but certainly not the final result, whereas the theoretical value proposed by the author as long ago as 1962 has not changed since then. Its accuracy is based on that of the universal quantum constants that he uses (c, G, h, e). (See Chapter VII – Chapter VIII).

In conclusion, the temporalistic model refutes the interpretation of the origin of redshifts as being the expansion of space, and interprets redshifts as being physical phenomena caused by the existence of the temporalistic constant, To, with a value of 4.5546×10^{17} s. The temporalistic model theoretically established, in 1962, the value of the temporalistic constant To as being 4.554610×10^{17} s. The cosmological data provided by WMAP 5, after 80 years of research by a large number of researchers, are approaching this value, within the range of uncertainties.

The evolution of galaxies – The large-scale structures of the Universe

The hierarchical model of creation and evolution of galaxies and large-scale structures in the standard Big Bang model raises a very large number of questions: what happened before the Planck time t (10^{-43} seconds)? What was the process of creation of matter? From nothing? How? What was the cause of the Big Bang? The answers to these questions are always either hidden from us or provided by new hypotheses.

The energy fluctuations that arose several thousand years after the Big Bang, from which galaxies are supposed to have formed through the action of gravity, are not enough to explain the evolution of large-scale structures Tegmark (2004).

In 2004, Brigitte Rocca revealed the existence of very young massive galaxies (at distances > 12 billion light years), which contradicts the hierarchical growth model (Dossier La Recherche 393 – January 2006).

The expanding Universe Big Bang model admits the existence of this repetitive yet irregular large-scale structure of the Universe, and especially of the huge voids measuring roughly 1×10^{26} cm to 1×10^{27} cm across. The standard model is unable to explain the causes of the existence of these huge voids, whose probability of existing is tiny (5×10^{-10}).

It may be considered that the large-scale structures of the Universe are made up of filaments made of gas, dust, stars, galaxies, clusters and superclusters of galaxies, great walls, great voids and dark matter.

The number of problems and inconsistencies in models of the growth of galaxies and large-scale structures is such that at present it can be considered that these problems have not been solved by the standard Big Bang model.

A detailed examination of a dozen of the most important concepts in the standard Big Bang model leads us to an inevitable conclusion: the so-called evidence and interpretations of the 12 concepts from the standard Big Bang model that we have analyzed all either show problems and inconsistencies, or are simple non-validated assertions, or are unverifiable speculations. In these conditions, there is very good reason to reject the standard model of cosmology, the Big Bang model, which amply justifies the title of our book:

The Big Bang Delusion

The standard Big Bang model has often been accepted by researchers <u>by default</u>, since no alternative model has been accepted. This option, which validates the standard Big Bang model, is not acceptable. It is counterproductive, since it validates an erroneous model for the simple reason that there currently exists no alternative. However, this does not mean that it is possible to accept a dogma that is manifestly wrong. The acceptation of an erroneous model compromises any proposal for a new model, for it is common knowledge that researchers who question the Big Bang model are not currently able to make their voices heard in the scientific organizations where astrophysics and astronomy are studied. A valid alternative model will be put forward sooner or later. This is our profound conviction, given that the Big Bang model is a model that is leading a considerable number of researchers in a direction that is both <u>a dead end and a waste of effort</u>. After all, the erroneous concept of the ether finally disappeared after decades of undisputed existence.

The author of this book has searched for an alternative model to a theory, the Big Bang model, which in his opinion is a delusion. The alternative model that he proposes is the temporalistic model. Time will tell whether this model is pertinent or not. In any case, even if it were not, the Big Bang dogma, given all the weaknesses and inconsistencies that we have pointed out, can in no way claim to put forward a credible model of the Universe, either past or present.

In Chapter XI, we showed how thinking about the nature of time and its fundamental asymmetry led us to the concept of the temporalistic constant To, with a value of 4.55465×10^{17} seconds (around 14.43 billion years) and to the temporalistic model that we developed ananthropically, that is, by validating it through observations and evidence, in accordance with the criteria that we consider to be fundamental, namely Popper's 'falsifiability' and Einstein's 'observable facts'. Our research naturally led us to the critical analysis of the standard model of cosmology, the Big Bang model, and to the inescable conclusion that it is a delusion.

<u>The Big Bang Delusion:</u> *Ad hoc* and 'off the peg' inflationary hypotheses, where "all that is required is to change a few parameters".

<u>The Big Bang Delusion:</u> The Big Bang singularity. In general relativity, singularities mark the boundary of the validity of this theory, while the many unification theories (superstrings, quantum gravity, non-

commutative geometry, etc) that claim to eliminate such singularities have failed (L'invention du Big Bang - Jean-Pierre Luminet).

The Big Bang Delusion: It is not possible to go further back than the Planck time t (10^{-43} seconds after the Big Bang), since the equations of both general relativity and of quantum field theory can no longer be used due to the appearance of many infinite terms.

The Big Bang Delusion: The cosmic microwave background interpreted as fossil radiation: the CMB had often been predicted, without using the Big Bang model, and long before Gamow, by: Guillaume (1896), Eddington (1926), Regener (1933), Nernst (1933), McKellar and Herzberg (1941), Finlay-Freundlich (1953) and Max Born (1953). These authors predicted temperatures ranging from 1.9 K to 6 K (André Koch Torre Assis and Marcos Cesar Danhoni Neves - 1995). Gamow's prediction in 1953 of a cosmic background radiation at a temperature of 7 kelvins was based on a fallacious mathematical argument (Weinberg 1980).

The Big Bang Delusion: The redshifts of distant galaxies are not evidence of the Big Bang model. This is merely one interpretation of cosmological observations. These observations, far from being evidence, are therefore merely simple hypotheses, interpreted in a way that is favorable to another hypothesis, the standard Big Bang model.

The Big Bang Delusion: The hypothesis of the existence in the Universe of a form of matter that has negative pressure brings us back to the cosmological constant Λ, considered to be similar to vacuum energy, which is known to result from the predictions of quantum field theory, but which leads to a totally unacceptable value, 60 to 120 times greater than the value inferred from cosmological observations.

The Big Bang Delusion: The critical analysis of the many problems raised by the standard Big Bang model (the horizon problem, the problem of flatness and critical density, the problem of the homogeneous, isotropic Universe, etc) implies that the only possible solution to these problems is the highly controversial hypothesis of inflation, as we saw earlier. Far from being a solution to these problems, the use of the concept of inflation does nothing more than pile up uncertainties on top of other uncertainties (new epicycles on top of other epicycles!)

The Big Bang Delusion: The 'primordial explosion' of the Universe, according to the Big Bang model, starting out from 'infinite' density and temperature, infringes all the current laws of physics. No *ex nihilo* creation of space, time, matter or energy has ever been observed, whether in physical or biological phenomena. No experimental or factual validation has ever been provided for this notion. The only 'evidence' consists of <u>mere assertions and/or hypotheses.</u>

This is not an exhaustive list of all the Big Bang delusions, and indeed it might well be more appropriate to speak about the <u>Big Bang Delusions than the Big Bang Delusion.</u>

What is extremely serious is that, in our opinion, current cosmology has been led into a dead end that leads nowhere, except to further speculations, erroneously validated by new, ever more risky hypotheses. For many years now, it has also resulted in <u>financial and above all human waste</u>, with thousands of researchers being guided towards fruitless research that leads nowhere.

In the last chapter of this book, we will compare the responses to cosmological problems provided by the Big Bang model and by the temporalistic model. The reader will be able to weigh up the pertinence of each of the two competing models.

PART SEVEN

Comparison between the Big Bang model and the temporalistic model.

CHAPTER XIV

Comparison – Conclusion - Tests

COMPARISON

We shall compare the arguments for the Big Bang model and the counter-arguments for the temporalistic model concerning the most important concepts that we have studied.

1) Redshifts.

The Big Bang model

Redshifts are caused by the expansion of space. Since the very first estimations of the Hubble constant in 1929 (Ho = 500 km/sec/Mpc and to = 2 billion years), the standard Big Bang model has, over the decades and after many adjustments and very numerous observations of redshifts, arrived at roughly the values established in 1962 by the temporalistic model, i.e. Ho = 67.71 km/s/Mpc and To = 14.43 billion years (4.5546×10^{17} s).

The Big Bang model is an interpretation (hypothesis) of redshifts based on cosmological observations. Redshifts are related to space. They make up the evidence and the premises of the Big Bang model, from which the entire theory results.

The temporalistic model

The concept of the expansion of physical space means that the vacuum is assigned properties which, by definition, it cannot have, such as curvature or speed. It is an inconsistent and therefore anthropic concept. If this interpretation is wrong, the entire Big Bang model collapses. Redshifts are caused by the existence of the temporalistic constant To.

The concept of the temporalistic constant To was drawn up <u>theoretically</u>. It is an interpretation (hypothesis) of cosmological observations. Its value was established by the author, in <u>1962</u>, at 4.5546×10^{17} seconds (around 14.43 billion years), and the value of the 'recession effect' of galaxies, Ho, at 67.71 km/s/Mpc. These values, established in a strictly theoretical way, were <u>validated</u>, after 80 years of cosmological observations, by NASA (WMAP5) in 2008, within the <u>range of uncertainties</u>. Redshifts are related to <u>time</u>.

<u>Moreover, on 25 July 2011, an Australian team, Florian Beutler et al (ICRAR et UWA), published new information: "The new measurement of the Hubble constant is 67.0 ± 3.2 km s^{-1} Mpc^{-1} ", which is very close to the temporalistic value 67.71 km/s/Mpc (to the nearest 1%) established theoretically in 1962 </u>by the author.

2) The Cosmic Microwave Background

The Big Bang model

The existence of the cosmic microwave background was predicted in 1940 by Ralph Alpher, Robert Herman and George Gamow as a consequence of the Big Bang model. They predicted it again in 1949. The cosmic microwave background predicted by Gamow and discovered by Arno Penzias and Robert Wilson in 1965 is fossil radiation dating from 380 000 years after the primordial explosion of the Big Bang. Tiny fluctuations of around 10^{-5} K in this radiation, which has a temperature of about 2.725 K, are observed. This is evidence for the Big Bang model. The anisotropies in the cosmic microwave background led to the formation of the first structures of galaxies.

The temporalistic model

Contrary to the historical information put around by the supporters of the Big Bang, the cosmic microwave background does not only result from the Big Bang model. It was predicted, without using the Big Bang model, and often well before Gamow, by Guillaume (1896), Eddington (1926), Regener (1933), Nernst (1933), McKellar and Herzberg (1941), Finlay-Freundlich (1953) and Max Born (1953). These authors predicted temperatures ranging from 1.9 K to 6 K (André Koch Torre Assis and Marcos Cesar Danhoni Neves - 1995). In addition, Gamow's prediction in 1953 of a cosmic microwave background at a temperature of 7 kelvins was based on a fallacious mathematical argument (Weinberg 1980).

The cosmic microwave background is not evidence for the existence of the Big Bang. Once again, it is merely an interpretation (i.e. a hypothesis) of a factual phenomenon, in correlation with another hypothesis, the Big Bang model. The cosmic microwave background is interpreted as being fossil radiation dating back 13.7 billion years. The small size of the fluctuations in the cosmic microwave background are not enough to justify quantitatively the origin and formation of the galaxies and the large-scale structures of the Universe (Tegmark). The cosmic microwave background that we observe is located at the temporalistic horizon, made up of the upper time limit To, which has a finite value of 4.5546×10^{17} s, or around 14.43 Ga. This is an observation and not an interpretation such as the 'fossil radiation' hypothesis in the Big Bang model.

The hypothesis of the cosmic microwave background, used as a validation of the hypothesis of the origin of the Big Bang, brings with it many problems, including the horizon problem, the fact that flatness of the Universe is almost the same as its critical density, and the existence of a homogeneous and isotropic Universe.

The standard Big Bang model cannot find a solution to these various problems. It is forced to call upon new, ad hoc, external hypotheses, the inflationary models, which have no experimental or factual support and which provide answers that are highly speculative.

The opinion of a renowned theoretical cosmologist and supporter 'by default' of the Big Bang, James Peebles, about the theory of inflation is edifying: "It is a theory that can be adjusted in order to produce the structures that we see, starting out from all the possible initial conditions. From this perspective, it is not really a theory but an 'off the peg' story, since it fits all cases. All that is necessary is to change a few parameters." (Les Dossiers de la Recherche – N° 35 – Quarterly - May 2009 – page 8).

To sum up, the explanation of the cosmic microwave background and the hypothetical inflationary theories connected to it are more similar to Ptolemaian epicycle-type reasoning than to a rigorous scientific model that respects current physical laws and principles (law of conservation of mass-energy, 'validated' or 'falsifiable' hypotheses, etc), i.e. ananthropic concepts and propositions. Far from supporting the standard model of cosmology, inflationary models aggravate its speculative nature.

3) Primordial nucleosynthesis

The Big Bang model

The agreement between the predictions of abundances of light nuclei using the basic hypotheses of the Big Bang and current abundances of these nuclei is one of the primordial nucleosynthesis model's strong points. However, it should be pointed out that there are a large number of versions of non-standard Big Bang scenarios. Thousands of articles have been devoted to them. They are based on initial conditions for the Big Bang that are different from the standard model (mainly the baryon/photon ratio, but also other hypotheses such as inhomogeneities, non-standard properties of neutrinos, etc). Nonetheless, all these models are based on the Big Bang model, but with different initial conditions.

All the hydrogen and part of the helium and lithium contained in the Universe were formed in the first hundred seconds after the Big Bang. Astrophysicists pay close attention to primordial nucleosynthesis. This is because the slightest result that refutes its predictions threatens the models of the Big Bang.

The temporalistic model

There is no primordial nucleosynthesis problem for the temporalistic model since, as far as this model is concerned, primordial nucleosynthesis doesn't exist. There was no 'primordial explosion'. The Universe exists. We have no evidence for its creation or for that of time, space or mass-energy. There is no evidence for its possible disappearance. The temporalistic Universe is a stationary Universe but one that is in constant dynamic evolution, whether photons or baryons, or stars, galaxies and large-scale structures.

According to the standard cosmological model, the value of baryonic density a few seconds after the Big Bang was between 3 and 5%. According to the map of fluctuations observed by the Boomerang collaboration (2000), the baryonic density in the cosmic microwave background dating from 380 000 years after the Big Bang was 7.4% (± 1 %). This discrepancy calls into question the standard Big Bang Nucleosynthesis model.

The origin of the creation of lithium a few minutes after the Big Bang is subject to debate. An unexpected origin for lithium has been discovered in red giants in around twelve globular star clusters. It may result from the decay of the unstable radioactive isotope beryllium-7. Lithium is also found in other very massive red giants at a late stage in their evolution (Catherine Pilachowski 2001). Moreover, the amount of lithium produced before the stars formed and the amount destroyed in stars is unknown.

A comparison of the results from the latest theoretical nucleosynthesis calculations and from WMAP 5 data show that the values inferred from the cosmic microwave background and astrophysical observations agree for deuterium, are no more than reasonable for helium-4, but are in complete disagreement for lithium-7.

All these discrepancies call into question the standard model of Big Bang nucleosynthesis.

The three pillars of the Big Bang model

The following three series of phenomena: 1) redshifts, 2) the cosmic microwave background, and 3) primordial nucleosynthesis, are generally considered to be the three key pillars that support the standard model of cosmology. Far from being evidence, they are merely hypotheses and interpretations. We can see that, according to our analysis, these three pillars, far from being set in stone, rest on quicksand.

The redshifts of distant galaxies are interpreted as being caused by the expansion of the Universe and by the recession of galaxies. Such concepts are not observational data. They result from an interpretation, i.e. a hypothesis based on confusion between the geometrical and physical concepts of spacetime. The temporalistic model proposes an alternative interpretation of redshifts, one that has been validated.

The interpretation of the cosmic microwave background as 'fossil radiation' is a hypothesis or an interpretation and not evidence for the Big Bang model, and far from supporting this theory, it piles up problems for the Big Bang concept, as we saw earlier: the problems of the horizon, the flatness of the Universe, critical density, the homogeneity and isotropy of the Universe, etc.

A recent observation is in serious disagreement with primordial nucleosynthesis (quasar APM 08279+5255, with an age of 13.5 billion years, contains 3 times more iron than the Solar System, which is around 5 billion years old - XMM-Newton - G. Hasinger and S. Komossa - July 2002).

The many difficulties for primordial nucleosynthesis that we mentioned earlier, including the major discrepancy of the abundance of lithium-7, are crippling for the validity of the primordial nucleosynthesis concept, a hypothesis that directly results from the Big Bang paradigm.

The 'THREE PILLARS' of the model, redshifts, the cosmic microwave background and primordial nucleosynthesis, which are, according to the supporters of the standard Big Bang model, irrefutable evidence, emerge much weakened from the critical analysis we have subjected them to. We have seen that the aforementioned three series of phenomena are mere hypotheses or interpretations, which are claimed as evidence when in fact they are nothing more than mere hypotheses or interpretations of existing facts.

4) The horizon problem

The Big Bang model

Observations of the cosmic microwave background show that, on large scales, the Universe is homogeneous and isotropic (with a precision in the region of 10^{-5}). Before inflation, the regions of the Universe which were still very close to each other had 'all the time in the world' to exchange their properties (such as temperature for instance). With inflation, these neighboring regions moved away from each other. Expansion was a local phenomenon which took place homogeneously at every point in the primordial Universe. This model is a representation in time and not in space. It is reasonable to suppose that, shortly after the Big Bang, all the matter observed was located in a small region, so that it can be assumed that this region was homogeneous and isotropic, and that the Universe then underwent a period of exponential expansion (inflation) which moved the different regions in this area away from each other very rapidly. However, it is very difficult to explain why, right from the start, the Universe ended up being homogeneous and isotropic.

The solution is inflation, which when it replaces normal expansion, enables an exponential expansion of the Universe to occur, without violating the speed limit of special relativity. This solution is possible, according to the Friedmann equations, by assuming that a form of matter that has negative pressure exists in the Universe.

The temporalistic model

The horizon problem does not arise in the temporalistic model. There is neither expansion nor inflation. The temporalistic horizon consists of the upper time limit, To, which has a finite value of 4.5546×10^{17} s.

In the Big Bang model, the horizon problem is only resolved by using a new hypothesis, that of inflation. The only theoretical justification for this highly speculative, *ad hoc* hypothesis, which is not based on any experimental evidence and relies on an exponential extrapolation of the laws of physics, is that it gets round the problems of the Big Bang model. Far from solving the horizon problem, this 'non falsifiable' hypothesis, whose many difficulties were analyzed earlier, is merely an additional new hypothesis. (See the harsh criticism of inflationary models by James Peebles, a supporter by default of the Big Bang model – paragraph 2).

The hypothesis of the existence in the Universe of a form of matter that has negative pressure brings us back to the cosmological constant Λ, considered similar to vacuum energy, which is known to result from the predictions of quantum field theory, leading to a totally unacceptable value, 60 to 120 times greater than the value inferred from cosmological observations.

5) The problem of flatness and critical density

The Big Bang model

Observations show that the Universe is almost completely flat, with an energy density of the same order of magnitude as the critical density corresponding to a universe with zero spatial curvature. Why should this be? The solution is the same paradigm that provides a solution to the horizon problem: inflation. If inflation increases the size of the Universe by a factor of 10^{50}, its curvature is reduced by an identical factor. Its current value is therefore very close to zero and its energy density very close to the critical density.

The temporalistic model

The problem of flatness and critical density does not arise in the temporalistic model. In the temporalistic model, space is a vacuum. By definition, it cannot be curved. Only its container, material space, can be curved. The concept of critical density has no reality in the temporalistic model. The energy density of the Universe, before and after inflation, could take any value. The Big Bang model provides no explanation for the flatness of the Universe.

Inflationary theories, which are highly speculative, are supposed to solve the flatness problem even though they are themselves a source of serious difficulties. The solution proposed for the flatness problem, which is identical to the solution to the horizon problem, namely inflation, therefore suffers from the same difficulties, in other words it is a highly speculative *ad hoc hypothesis*, with the concept of matter with negative pressure, lacking any observational backing, and with a totally unacceptable value, 60 to 120 times greater than the value inferred from cosmological observations

(See the harsh criticism of inflationary models by James Peebles, a supporter by default of the Big Bang model – paragraph 2). According to James Peebles, an eminent supporter of the Big Bang model as we saw earlier, a major argument that leads to the rejection of inflationary theories is their ability to fit all possible initial conditions (Chapter VIII: Inflationary theories, page 83).

6) the problem of the homogeneous and isotropic Universe

The Big Bang model

Observations show that the Universe is homogeneous and isotropic. The COBE satellite, launched in 1989, confirmed that the temperature of the cosmic microwave background (approximately 2.73 kelvins) is isotropic, i.e. identical in all directions, varying by less than one part in a hundred thousand.

It can be shown, using the Friedmann equations, that a homogeneous and isotropic universe will remain in this state. However, it is hard to argue that this homogeneous and isotropic state of the Universe was <u>originally</u> thus so. The solution is the same paradigm that provides a solution to the horizon problem: inflation. The parts of the Universe that are observable today were causally linked before inflation. After inflation, the size of the Universe had increased by 10^{50}, and the result is homogeneous and isotropic radiation in every region of the Universe.

The temporalistic model

For the temporalistic model, which considers that cosmological observations are contradictory, a 'homogeneous and isotropic Universe', as asserted by the Big Bang model, is a <u>mere hypothesis</u>, without any validation.

<u>Theoretically, there is no evidence or valid reason to assume the existence of a homogeneous and isotropic Universe right from its origin.</u> There is also no valid explanation for anisotropies in the cosmic microwave background of around one part in a hundred thousand. The solution proposed is always the same paradigm, inflation.

Inflationary theories are supposed to solve the problem of a homogeneous and isotropic Universe, like the horizon and flatness problems, even though they are themselves a source of serious difficulties.

All we can do is to repeat the arguments put forward in the final two paragraphs of the previous chapter, which totally invalidate inflationary theories.

According to the cosmological principle, the Universe is homogeneous and isotropic. <u>This is a mere hypothesis without any validation.</u> The various

inhomogeneous structures already discovered seriously challenge the standard model of cosmology, based on the cosmological principle, which gives the Universe a homogeneous and isotropic structure. And yet the Universe does not seem to be in the least uniform either on small or large scales. The Universe appears to be made up of filaments where clusters, super clusters and hyperclusters of galaxies, and large-scale structures such as walls and great voids, are collected together (Rudnick). The expanding Universe Big Bang model admits the existence of this repetitive yet irregular large-scale structure of the Universe, and especially of the huge voids measuring roughly 1×10^{26} cm to 1×10^{27} cm across. <u>The standard model is unable to explain the causes of the existence of these huge voids, whose probability of existing is tiny (5×10^{-10}).</u>
On the other hand, the temporalistic model offers a simple explanation for the structure of the Universe and for the existence of filaments and great voids. In the temporalistic model, gravitation has a finite range, embodied by the concept of gravitational radius $r = m^{1/2}$ (r= radius, m = mass). In the filaments and at their junctions, the gravitational effect of galaxies and galaxy clusters operates lengthways, since the masses are relatively close and therefore below the threshold of the gravitational radii. If we take the example of a rich galaxy cluster (3 000 galaxies) whose mean mass is around 1×10^{49} g, its gravitational radius is $(1 \times 10^{49})^{1/2}$ cm = 3×10^{24} cm. It can therefore have a gravitational influence on galaxies and galaxy clusters whose average distance is 1 Mpc (3×10^{24} cm) (See Chapter XII – Temporalistic gravitation – Masses and gravitational radius, paragraph 10).

7) The origin of the Big Bang

The Big Bang model

The expanding Big Bang universe was originally a spatio-temporal singularity which must have been infinitely dense. According to the theory of inflation, the visible Universe originated in a very small, very hot (10^{32} kelvins) region of the homogeneous Universe which inflated 10^{-35} seconds after the Big Bang. This inflationary phase lasted 10^{-32} seconds during which the Universe expanded by a factor of around 10^{50}, after which the Big Bang continued to evolve.

Observations of the Universe that are accessible to telescopes are located 380 000 years after the Big Bang, in other words when the cosmic microwave background radiation was emitted. Moreover, it is not possible to go further back than the Planck time (10^{-43} seconds after the Big Bang), since the equations of both general relativity and of quantum field theory can no longer be used due to the appearance of many infinite terms. The latest data provided by WMAP 5 (Table 7 – Cosmological Parameter Summary – 2008) gives a value for Ho = 71.9 (+2.6 – 2.7) km/s/Mpc and to = 13.69 (± 0.13) billion years.

The Big Bang brings about the appearance of space and time, or of spacetime, as well as of matter and energy. Since time was created at the same time as the Big Bang, it is impossible to go back any further, in other words beyond 13.7 billion years.

The temporalistic model

According to the Big Bang model, the Universe was born, in the 'primordial explosion', from a spacetime singularity which had an 'infinite' density and temperature. What was the cause of this explosion? No answer to this question is provided by the current laws of physics. Or else, this difficulty is side-stepped by denying the 'primordial explosion', without any clear or valid justification. Where did space, time, matter and energy come from? They were created *ex nihilo*, equally without any experimental or factual validation. No *ex nihilo* creation of matter or energy has ever been observed, whether in physical or biological phenomena.

To assert that space and time appeared with the Big Bang is a circular argument which obviously eliminates, arbitrarily and without any validation, the problem of the existence of time before the Big Bang.

It is also equally possible to assert, without any other justification, that at the singularity of the Big Bang, the notion of space disappears but not that of time (Gabriele Veneziano's pre-Big Bang). As usual, this new hypothesis is not 'falsifiable'.

Many other hypotheses have been put forward: the ekpyrotic model, which proposes a multidimensional brane universe where inflation is replaced by the cyclic collision between two universes; the strictly speculative model of eternal inflation of bubble-Universes, for which there is no factual or experimental evidence, etc, etc.

All these models are strictly speculative, without there being any possibility of validating them. However, this does not appear to bother their authors, who claim the right to speculate without any restrictive tests of validity.

When all is said and done, the Big Bang model is a strictly anthropic concept. It infringes several of the ananthropic criteria that we set out in Chapter III: it is irrational and speculative at the expense of critical thinking; it infringes current physical laws without providing any experimental or observational validation; it uses contradictory concepts such as 'quantum vacuum filled with quantum fluctuations', etc.

Refuting the Big Bang and its highly speculative origins, the temporalistic Universe does not suffer from the problems that we have just mentioned. The temporalistic Universe is a stationary but evolving Universe. There was no primordial explosion; no 'expansion of space'; no singularity with infinite physical values; and no inflation. The temporalistic model assumes nothing. It simply observes the existence, in the Universe, of space, i.e. a vacuum, and of time, i.e. the various durations of phenomena, matter and energy.

8) Inflationary theories

The Big Bang model

The difficulties encountered by the Big Bang model, the horizon problem, the fact that the flatness of the Universe is almost the same as its critical density, and a homogeneous and isotropic Universe, have led theoreticians to seek for new approaches that might get round such problems.

This is what led to the creation of *ad hoc* and entirely speculative hypotheses, the inflationary theories, which have come in various versions: the theory of inflation elaborated by Alexei Starobinsky was developed by Allan H. Guth and Paul Steinhardt (1984 – 1998), Andy Albrecht, and Andrei Linde (1994 – 2001).

The theory of inflation has the virtue of solving a certain number of problems raised by the Big Bang model:

1) The extraordinarily rapid inflation of the Universe, at speeds well over the speed of light, originating in a tiny, homogeneous region of the Universe solves the horizon problem.

2) A flat Universe with a density close to the critical density results from the inflationary model.

3) The problem of magnetic monopoles: for the creation of nuclei in the primordial Universe, the Big Bang model requires the use of the Grand Unification Theory (GUT) and the production of massive particles, called magnetic monopoles. Many of these magnetic monopoles should still be around today. The lack of any magnetic monopoles today is explained by their rapid dispersal during the inflationary phase.

The inflationary model predicts small fluctuations in the cosmic microwave background, leading to the formation of galaxies.

The temporalistic model

The theory of inflation is an extension of the Big Bang model, but it is independent of it.

The inflationary model, created in order to solve the problems of the Big Bang model, is not based on any experimental or factual evidence. The considerable extrapolation of the laws of physics set out in this model has no theoretical justification, apart from providing an arbitrary response to the difficulties of the Big Bang model. In the final analysis, it is nothing more than an _ad hoc hypothesis_. "The assertions of the inflationary model, which are poorly demonstrated, can lead to genuine scepticism in the eyes of rigorous observers" (Peebles 2001). Incidentally, piling up hypotheses without any observational basis, can lead to highly speculative and particularly questionable versions of inflation: chaotic inflation, self-reproducing universes, multiple universes, parallel universes, bubble universes with eternal inflation, creation of universes in a laboratory, creation of universes by a physicist-hacker, and other wild ideas light years away from the required scientific rigor! There is nothing to stop physicists' imaginations from running away with themselves since any experimental link with reality has disappeared!

To get round the various difficulties of the Big Bang model, cosmologists have transferred them to another, even more hypothetical and questionable concept, inflation. In fact, this is to jump out of the frying pan into the fire.

The cause of inflation, which began when three out of the four fundamental interactions had dissociated, remains unknown. The beginning and then the end of inflation are only justified by means of further hypotheses. The existence of the cosmological constant Λ, rejected by Einstein, which is required by inflationary models, at present remains a pure hypothesis which leads to insurmountable difficulties with regard to physical reality, with a totally unacceptable value, 60 to 120 times greater than the value inferred from cosmological observations

Another major argument that leads to the rejection of inflationary theories is their ability to fit all possible initial conditions, which James Peebles, a renowned cosmologist who supports the Big Bang model 'by default', was quick to point out: "It (inflation) is a theory that can be adjusted in order to produce the structures that we see, starting out from all the possible initial conditions. From this perspective, it is not really a theory but an 'off the peg' story, since it fits all cases. All that is necessary is to change a few parameters." (Les Dossiers de la Recherche – N° 35 – Quarterly - May 2009 – page 8).

The highly speculative concept of inflation has no place in the temporalistic model, which rejects all speculative and unverifiable concepts, in other words, anthropic concepts. The temporalistic model is thus unaffected by any of the problems pertaining to inflationary theories.

9) The acceleration of expansion – Dark energy

The Big Bang model

The hypothesis of accelerating expansion leads to the hypothesis of the existence of dark energy. Evidence for the acceleration of the expansion of the Universe: Type Ia supernovae, galaxy cluster counts, gravitational lensing, and evidence for the existence of dark energy: type Ia supernovae, the cosmic microwave background (and its fluctuations) directly correlated to the geometry of the Universe (flat according to Boomerang), followed by WMAP 5 and acoustic waves.

Models for the nature of dark energy: a) the cosmological constant, Λ, likened to vacuum energy, according to the predictions of quantum field theory, with quantum vacuum fluctuations. In quantum field theory, the vacuum is not nothingness, but rather the fundamental minimum energy state of quantum field systems; b) quintessence; c) modified general relativity; d) axions, the transformation of some photons into axions which go undetected by telescopes, which means that the luminosity of galaxies is underestimated, which is interpreted as evidence of accelerating expansion. Dark energy was introduced into the standard Big Bang model in the form of a cosmological constant Λ.

The temporalistic model

"Type Ia supernovae are used to evaluate the expansion of the Universe assuming a standard mechanism of formation. However, there is nothing standard about the formation of type Ia supernovae (supernova SN2006gz), and this distorts cosmologists' measurements (Stéphane Fay – Astrophysical Journal Letters, vol. 669 pp.L17-L19.2007). "Several teams have shown that certain models without accelerated expansion could reproduce the observations of supernovae if it is assumed that we live in a reduced density region of the Universe, a sort of bubble whose density is lower", Jean-Philippe Uzan (Dossiers La Recherche - May 2009 – p 91).

The acceleration of expansion leads to the hypothesis of the existence of dark energy. Various models propose an explanation for this: a) the cosmological constant Λ, introduced by Einstein and then rejected by him (according to Einstein, it was the greatest blunder in his life), and considered similar to vacuum energy. Unfortunately, the predictions of quantum field theory give a <u>totally unacceptable value, 60 to 120 times greater</u> than the value inferred from cosmological observations. This value, inferred from quantum field theory and incompatible with the properties of the Universe, constitutes a <u>major conceptual problem that is still unresolved</u>; b) quintessence (very popular a few years ago, but abandoned since because of the many problems it raises; c) general relativity with its 'scalar tensors'; no observation has validated this concept, which remains a pure hypothesis; d) axions, particles that result from the transformation of a certain proportion of photons: this model has today been abandoned. Many of these models (such as quintessence) have 'free functions' that can be fitted to those of the cosmological constant Λ, thus making them impossible to refute and therefore not 'falsifiable'. None of the models proposed has therefore been validated. In desperation, some have had no hesitation in proposing an 'anthropic model'!

Only inhomogeneous and non-isotropic models of the Universe, with their questioning of the cosmological principle, are free from this critique. They lead to a rejection of the acceleration of expansion and its consequence, the existence of dark energy.

According to the latest research by Arman Shafieloo and colleagues (14 April 2009), concerning nearby supernovae (less than a billion light years away), the acceleration of expansion has diminished over the last 2.5 billion years, to the point of reversing recently. This implies a similar fall in the density of dark energy, which would mean the exclusion of the cosmological 'constant' Λ (http://arxiv.org/abs/0903.5141).

To sum up: the concepts of expansion, accelerating expansion and dark energy, with all the problems that they lead to, are hypotheses that are absent from the temporalistic model. They are the direct result of the spatial interpretation of the redshifts of distant galaxies. The temporalistic (in other words, temporal) interpretation of redshifts avoids all these problems because it is based on the geometry of a Universe that is inhomogeneous on large scales. According to the temporalistic model, the Universe is stationary and evolving. In it, there is no expansion, no inflation, no accelerating expansion and, *a fortiori*, no dark energy.

10) The theoretical prediction of the Hubble constant, Ho – The age of the Universe, to

The Big Bang model

Today, Hubble's law is interpreted not as being caused by the motion of galaxies through space, but rather by the expansion of space itself (within the framework of general relativity rather than special relativity, since the latter prohibits speeds faster than the speed of light, c). In 1929, the value of the Hubble constant was estimated to be around 500 km/s/Mpc, due to a wrong estimation of the absolute magnitude of the Cepheids. The latest data provided by WMAP 5 (Table 7 – Cosmological Parameter Summary – 2008) gives a value for <u>Ho = 71.9 (+2.6 – 2.7) km/s/Mpc and to = 13.69 (± 0.13) billion years.</u>

The age of the Universe represents the time that has passed since the Big Bang, i.e. the dense, hot phase of the Universe.

When the acceleration of the recession velocity of galaxies is constant, it can be found using many different methods: Cepheids, type Ia and type IIa supernovae, the study of the fundamental plane of the galaxies, and shifts in fluctuations of brightness of the multiple images of quasars produced by gravitational lensing effects.

The age of the Universe, to, = 1 / Ho if the Universe has a very low matter density, which is what observations show (an almost flat Universe).

Corrections can be made to Hubble's law.

General relativity and the Friedmann-Lemaître equations lead to a change in the scale factor R(t) as a function of time t, the expansion of space implying that R(t) is becoming greater.

The current expansion rate Ho is today estimated to be 10 times lower (70 km/s/Mpc), that is, $1 / Ho = 14 \times 10^9$ years. The theory's other free parameters (the density parameter for the Universe and the cosmological constant Λ) have begun to be determined observationally since 1998. They cancel each other out, giving an age close to 1 / Ho. In 2008, the value of to in the 'concordance' model was estimated to be between 13.7 and 13.8 billion years.

The temporalistic model

The latest estimations of the value of the Hubble constant, Ho, and of the age of the Universe to cited above are the result of 80 years of observational research and successive approximations. Over the decades, the value for Ho has changed from 625 km/s/Mpc to 71.9 km/s/Mpc (+ 2.6 – 2.7) and for to from 1.6 billion years to 13.69 (± 0.13) billion years. In his temporalistic model, the author established, <u>in a strictly theoretical manner, in 1962,</u> a value for the Hubble constant, Ho, of 67.71 km/s/Mpc and for to (which he called the 'temporalistic constant, To') of 4.5546×10^{17} s, that is, around 14.43 billion years.

In Chapters VII and VIII we compared the observational value and the theoretical value of Ho: 69.2 km/s/Mpc (71.9 − 2.7) for the former and 67.71 km/s/Mpc for the latter, i.e. a difference of 2.16%. This difference is negligible if we consider the uncertainty in the WMAP 5 data: between 3.2% (+2.6) and 3.75% (-2.7). The value of the Ho constant theoretically established and proposed by the author is very accurate since it is based on the accuracy of the universal and quantum constants that he uses (c, G, h, e).

In conclusion, the temporalistic model refutes the interpretation of the origin of redshifts as being the expansion of space, and interprets redshifts (the increase in wavelength of moving photons) as being physical phenomena caused by the existence of the temporalistic constant, To, with a value of 4.5546×10^{17} s. Redshifts do not have a spatial meaning (the expansion of space in the Big Bang theory), but rather a temporalistic meaning (temporal effect of the temporalistic constant To on the wavelength – or energy - of photons in motion). In other words, the cause of redshifts is of a temporal nature rather than a spatial nature. Redshifts result from the nature of photons, which are affected, as they travel through space, by the existence of the 'temporalistic constant' To whose value is 4.5546×10^{17} s. This alteration of the energy of photons is in no way connected with the concepts of 'tired light' or of interaction with other physical particles (such as the Compton effect).

If the Universe has a very low matter density, which is the case, the 'age' of the Universe, to, equals $1 / Ho = 1 / 2.243 \times 10^{18}$ s $= 4.458 \times 10^{17}$ s, which is around 14.12 billion years. The differences with the values obtained by the author are, as for the values of Ho, in the region of 2.15% (Ho = 67.71 km/s/Mpc and To = 4.5546×10^{17} s), in other words within the range of the uncertainties.

The value of the Hubble constant established theoretically by the author in 1962, as well as that of To, the temporalistic constant (1/Ho) is not a new hypothesis but the direct consequence of the temporalistic (i.e. temporal interpretation) of the redshifts of distant galaxies. In fact, in the temporalistic theory, the temporalistic constant, To, is not the age of the Universe but rather a temporal constant, i.e. the duration of a phenomenon, the loss of energy by photons as they travel through space (See Chapter III – Anthropic and ananthropic concepts – b) The physical concept of time).

11) The evolution of galaxies – The large-scale structures of the Universe

The Big Bang model

Most scenarios for the formation of galaxies and large-scale structures currently favor the hierarchical model, in which structures form by successive mergers of subsystems.

Nonetheless, there are doubts about the scenario of hierarchical formation of galaxies since the Big Bang. According to the statistics that have been compiled about galaxies, they only really differ in terms of their mass. The accretion of gas may be the principal factor in the growth of galaxies (Pour la Science – N° 374 – December 2008 p 9). According to a new scenario for galaxy formation (unlike the standard model of formation by collisions of galaxies), galaxies form from currents of cold gas (Nature 2009 - Pour La Science March 2009 – N° 377 p 11).

According to the Big Bang model, what do the fluctuations in the cosmic microwave background represent? They are the actual record of the fluctuations that gave rise to the galaxies and to large-scale structures. The recombination of matter took place about 380 000 years after the Big Bang. According to WMAP5, the concordance model shows that: the age of the Universe is 13.7 Ga; it is made up of around 70% dark energy and 30% matter, of which 5% is ordinary (baryonic) matter and 25% dark matter. The model which best fits observations is the Lambda-Cold Dark Matter model (ΛCDM).

By using Hubble's law of expansion, the distances of fairly distant galaxies have been accurately determined.
When simulations of the formation of structures in a ΛCDM dark matter universe are compared with observations, three unresolved problems arise: 1) the radial distribution of dark matter in galaxies does not correspond to that inferred from their rotation curves; one possible solution is to 1) modify the law of Newtonian dynamics at low accelerations (Milgrom 1984); 2) "At equilibrium, the disks of spiral galaxies in simulations are ten times too small compared to observations"; 3) "the ΛCDM model predicts that all spiral galaxies like the Milky Way should be surrounded by at least 400 satellite galaxies, or 400 small dwarf galaxies." But according to observations, there are at most a mere dozen or so dwarf companions. What are the solutions?" (Grandes structures de l'univers - Françoise Combes – Astronomie, May 2005)

The temporalistic model

The model of creation and evolution of galaxies and large-scale structures in the Big Bang model raises a great number of problems: what happened before the Planck time (10^{-43} seconds)? What was the process of creation of matter? From nothing? How? What was the cause of the Big Bang? The redshift of distant galaxies, revealed by Hubble, on which the standard model of cosmology is based, implies a singularity with temperature, density and energy parameters with exceptionally high values. This singularity cannot be incorporated into current physics since the equations of both general relativity and of quantum field theory can no longer be used due to the appearance of many infinite terms (see The origin of the Big Bang - Chapter VIII).

"Why do we see some spiral galaxies, which are highly evolved structures, just a few billion years after the Big Bang?" (Françoise Combes)

"Theory says that elliptical galaxies could only have formed fairly recently. However, observation reveals elliptical galaxies that are already very old. Where is the mistake?" (James Peebles – Le Big Bang – La Recherche N° 35 – Quarterly - May 2009 p. 9)

The energy fluctuations that arose several thousand years after the Big Bang, from which galaxies are supposed to have formed through the action of gravity, are not large enough to explain the evolution of large-scale structures. According to Tegmark (2004), although the anisotropies in the cosmic microwave background are entirely in agreement with this idea on small and medium scales, this is not true at all on large scales. The way in which the structures develop depends on the origin of the primordial fluctuations and on the nature of dark matter.

In 2004, Brigitte Rocca revealed the existence of very young massive galaxies (at distances > 12 Gly), which contradicts the hierarchical growth model (Dossier La Recherche 393 – January 2006).

On the other hand, the temporalistic model offers a simple explanation for the structure of the Universe and for the existence of filaments and great voids. In the temporalistic model, gravitation has a finite range, embodied by the concept of gravitational radius $r = m^{1/2}$ (r= radius, m = mass). In the filaments, the gravitational effect of galaxies and galaxy clusters operates lengthways, since the masses are relatively close and therefore below the threshold of the gravitational radii. If we take the example of a rich galaxy cluster (3 000 galaxies) whose mean mass is around 1×10^{49} g, its gravitational radius is $(1 \times 10^{49})^{\frac{1}{2}}$ cm $= 3 \times 10^{24}$ cm. It can therefore have a gravitational influence on galaxies and galaxy clusters whose average distance is 1 Mpc (3×10^{24} cm) (See Chapter XII –Temporalistic gravitation – Masses and gravitational radius),all the way along the filaments.

Great voids (such as Rudnick's void measuring 1×10^{27} cm), as well as the various inhomogeneous structures already discovered, seriously challenge the standard model of cosmology, based on the cosmological principle, which gives the Universe a homogeneous and isotropic structure. The expanding Universe Big Bang model admits the existence of this repetitive yet irregular large-scale structure of the Universe, and especially of huge voids measuring roughly 1×10^{26} cm to 1×10^{27} cm across. The standard model is unable to explain the causes of the existence of these huge voids, whose probability of existing is tiny (5×10^{-10}). The huge sizes of the masses necessary for galaxies and galaxy clusters to have a gravitational effect on great voids, and the scarcity of such concentrations of galaxies, explains the existence of these voids, which is one of the serious challenges to the Big Bang model.

In the temporalistic model, there are no constraints on the evolution of galaxies and on the large-scale structures of the Universe: there are no primordial galaxies; no temporal priority for elliptical or spiral galaxies; no need for more or less large fluctuations in the cosmic microwave background causing small-scale and large-scale structures; and the model predicts that the structure of the Universe contains filaments, galaxy clusters and superclusters, dust, great walls and great voids, due to the fact that masses have gravitational radii with a finite range, which results directly from temporalistic gravitation. In the next paragraph, we shall examine the problem of dark matter, whose existence in the large-scale structures of the Universe plays a fundamental role.

12) Dark matter –The Pioneer effect – The MOND theory – The Casimir effect

Dark matter

The Big Bang model

Dark matter (or missing matter) is estimated to be around 80-90 % of all matter. It reveals its presence not only in galaxies but also in the large-scale structures of the Universe, and in galaxy clusters and superclusters. Many candidates have been proposed (MACHOs, neutrinos, WIMPs, brown dwarfs, supermassive black holes, etc) but, for now, its nature remains unknown.

What is currently known about the nature of dark matter?

1) The mass/luminosity relation, according to the distance, confirms the existence of an invisible type of matter, not only around galaxies but also between them.

2) The rotation curve (velocity) of galaxies allows us to conclude that stars and other luminous bodies make up less than 10% of the total mass of a galaxy. The remaining 90% is made up of dark matter or is under the influence of an unknown phenomenon.

3) The rotation curves of galaxies suggest that dark matter is contained in vast halos surrounding the galaxy's visible stars.

4) It is impossible to find dark matter far from galaxies, in very extensive halos, because tidal forces would disperse it throughout the entire cluster containing the galaxies.

5) Using the gravitational lensing method to study the effects of dark matter on the galaxy cluster Abell 1689 (distortion depending on the mass and radius of the deflecting galaxies), as proposed by the physicist Anthony Tyson, shows that "dark matter makes up over 90% of all matter".

6) To a large extent, dark matter accompanies luminous matter wherever it is located in galaxies, galaxy clusters and even large-scale structures tens of megaparsecs across.

7) Dark matter accompanies the irregularities in the distribution density of luminous matter throughout the visible Universe.

8) Dark matter does not exist or does not reveal its presence in the great voids that are tens or hundreds of megaparsecs in size (Richard Schaeffer 2001).

The temporalistic model

Because of the large number of unexplained cosmological observations, the Big Bang model postulates the existence of dark matter and describes its possible main characteristics. However, it is unable to specify either its nature (MACHOs, WIMPS, dwarf stars, etc?) or its origin. The concept of dark matter has no connection with that of the Big Bang.

The existence of dark matter is not a hypothesis, but an <u>inevitable</u> consequence of the temporalistic model. According to this model, the luminous matter in the Universe, in other words all the photons that are emitted in the radiation of every luminous source in the Universe, lose energy (redshift) as they travel, which is what causes the temporalistic acceleration field. This field, which is made up of gravitons, is what causes the universal gravitational field, which has a finite range (see Chapter XII – Temporalistic gravitation – Masses and gravitational radius).

The characteristics of dark matter fit the description given above, especially paragraphs 1: presence around galaxies and also between them; 3: presence in the vast haloes that surround the visible stars; 4: it is impossible to find dark matter far away from galaxies; 6: dark matter generally accompanies luminous matter; 7: dark matter accompanies the irregularities in the distribution density of luminous matter; 8: dark matter does not exist or is not observed in great voids. This last point, (8), confirms, *a contrario*, the <u>absence of dark matter where there are no stars</u>.

Independently of the qualitative arguments of the temporalistic model for dark matter, an irrefutable quantitative piece of evidence is provided by the quantitative value of the rotation curve of galaxies and the value of the acceleration of the velocity of stars in galaxies, attributed to the effect of dark matter and established by cosmological observations, <u>which is indeed of the order of the value of the temporalistic gravitational constant G', 6.582×10^{-8} cm/s².</u>

Very recent observations (Benoit Famaey and colleagues – Strasbourg Observatory - G. Gentile et al. Nature, 461, 627-628, 2009) confirm the correlation between luminous matter and dark matter: "Astonishing relationships thus appeared: from one galaxy to another, the strength of gravity caused by dark matter at the characteristic radius is identical, and the same applies to the strength of gravity caused by visible matter at the same radius. What can be inferred from these relationships? Firstly, that there is an inverse correlation between the central density of dark matter and that of visible matter. A high central density of visible matter implies that the density of dark matter at the center is low, and vice versa. Secondly, that the ratio of the densities of visible matter and dark matter which applies over the whole Universe remains valid within the characteristic radius for all galaxies."

These observations are consistent with the temporalistic model, according to which dark matter results from luminous sources.

The anomalous radial acceleration of Pioneer 10

For over 20 years, a problem has intrigued planetary scientists and physicists: "a slight, unexplained sunward acceleration of the motion of the Pioneer 10, Pioneer 11 and Ulysses spacecraft" (www.geocities.com/solarstormmonitor/Pioneer.html). Many other websites provide information on this subject.

The anomalous acceleration has several characteristics:

1) Its value, according to different authors, is 7.59×10^{-8} cm/s^2 (http://renshaw.teleinc.com/papers/prl-pi/prl-pi.stm),
8.74 ($\pm$ 1.33) $\times$ 10^{-8} cm/s² (http://csep10.phys.utk.edu/newsgroups/mond/messages/22.html),
or "around 10 billion times smaller than the acceleration we feel, the Earth's gravitational attraction" (www.geocities.com/solarstormmonitor/Pioneer.html, http:).//spaceprojects.arc.nasa.gov/Space_Projects/pioneer/PNStat.html).

2) The order of magnitude of this anomalous acceleration is $c \times Ho$ (Hubble constant).
3) This anomalous acceleration, which is independent of distance, is constant with regard to the speed of the spacecraft.

4) This anomalous acceleration is radial.

The Big Bang model

The Big Bang model provides no explanation of the Pioneer 10 effect.

The temporalistic model

The description of the characteristics of the anomalous radial acceleration of Pioneer 10 corresponds to that of the temporalistic gravity field.

When spacecraft leave a circular or elliptical path and take on a radial path leading out of the Solar System, the radial effect of the temporalistic universal acceleration field makes itself felt, and reduces the speed of the spacecraft (Pioneer 10, Pioneer 11, Ulysses, Galileo, etc.).

The temporalistic universal acceleration field does not affect the circular or elliptical orbits of the planets in the Solar System but only radial paths.

Moreover, this unexplained effect confirms very precisely the value of the temporalistic universal, isotropic acceleration field G', that is, G' = c / To, where G' is the temporalistic gravitational constant, c the speed of light, and To the temporalistic constant, that is G' = 2.997925×10^{10} cm/s / 4.5546×10^{17} s = $\underline{6.582 \times 10^{-8} \text{ cm/s}^2}$.

Observational values of the acceleration differ slightly from the value of the temporalistic gravitational constant G' (7.59×10^{-8} cm/s² and $8.74 (\pm 1.33) \times 10^{-8}$ cm/s² as compared to 6.582×10^{-8} cm/s²). This slight difference is due to the greater accuracy of the theoretical value of the temporalistic gravitational constant G'.

An experimental measurement validates the temporalistic model. By September 1998, the slowing down of Pioneer 10 had caused a reduction in the distance covered, compared to its predicted path, of some $\underline{400\ 000 \text{ km}}$. The radial journey of Pioneer 10, which began in 1973 – 1974, had therefore lasted some 24.5 years, i.e. 7.73×10^8 s. During this period, the deceleration, given an acceleration constant of 6.582×10^{-8} cm/s², was equal to 6.582×10^{-8} cm/s$^2 \times 7.73 \times 10^8$ s $\times 7.73 \times 10^8$ s $= 3.93293 \times 10^{10}$ cm = <u>393 293 km</u>.

The MOND theory

The MOND theory proposes that when the acceleration inferred from the Newtonian acceleration constant Gn is less than a°, that is Gn < a°, Newtonian theory does not apply, the parameter a° being comparable in value to c × Ho. The MOND theory is proposed as an alternative to dark matter.

The Big Bang model

The Big Bang model does not subscribe to this theory which does away with the concept of dark matter.

The temporalistic model

The temporalistic model does not deny the existence of dark matter. Quite the opposite. According to the temporalistic model where Ho = 1 /To, a° ~ c × Ho = c / To, that is 6.582×10^{-8} cm/s^2. When the acceleration caused by a mass is less than G', the Newtonian model no longer applies in MOND theory. In the temporalistic model, Newtonian theory no longer applies for an acceleration smaller than G', as in MOND theory, but this is due to the <u>finite graviational radius of masses and to the temporalistic universal acceleration field G'</u>.

The Casimir effect

The Casimir effect, named after its discoverer, is an effect that exists between two parallel conducting metal plates which, when placed very close to each other, attract each other.

The Big Bang model

This force is supposed to result from the concept of the quantum vacuum, which is not really a vacuum but the seat of fluctuations which create virtual particles that exert an attractive force on the plates.

The temporalistic model

The temporalistic model proposes an alternative to the quantum explanation.

The isotropic acceleration field of value G' created by the energy loss of photons is disturbed by the presence of the two metal plates. The result is that there is a smaller acceleration force between the two plates, causing them to move closer to each other. It would be interesting to calculate whether this temporalistic effect is confirmed quantitatively.

Summary of the comparison between the Big Bang model and the temporalistic model.

The Big Bang model

Number of hypotheses and interpretations: 13 (of which 11 are hypotheses and 2 are interpretations): 1 – interpretation of redshifts as being caused by the expansion of space; 2 – interpretation of the cosmic microwave background as being fossil radiation; 3 – hypothesis of primordial nucleosynthesis; 4 – solution to the horizon problem by the inflation hypothesis; 5 – solution to the flatness and critical density problems by the inflation hypothesis; 6 – solution to the problem of the homogeneous and isotropic Universe by the inflation hypothesis; 7 – the hypothesis of the origin of the Big Bang in the primordial explosion; 8 – the hypothesis of inflation and of the origin of space, time, matter and energy; 9 – the hypothesis of the accelerating expansion of space and of dark energy; 10 –

the hypothesis of the Hubble constant Ho and the age of the Universe To; 11 – the hypothesis of the of the origin of the galaxies and the large-scale structures of the Universe in the anisotropies of the cosmic microwave background; 12 – the hypothesis of dark matter; 13 – the hypothesis of the Casimir effect.

The Big Bang model says nothing about the Pioneer 1 effect and MOND theory. It provides no evidence for these 13 concepts.

The temporalistic model

Number of hypotheses and interpretations: 3 (of which 2 are interpretations and one is a hypothesis): 1 – interpretation of redshifts as being a temporal phenomenon caused by the existence of the temporalistic constant To and the temporalistic gravitational constant G'; concepts 2, 3, 4, 5, 6, 7, 8, 9 and 11 in the Big Bang model do not exist in the temporalistic model; 12 – interpretation of the temporalistic acceleration field as being dark matter.

Concept 10 (hypothesis of the Hubble constant Ho and the age of the Universe To) results from concept 1; concept 12 (interpretation of the temporalistic acceleration field as being dark matter) is validated by the agreement in the values of the temporalistic constant G' (6.582×10^{-8} cm/s²) and the values of the anomalous radial acceleration of Pioneer 10, and also those of MOND theory; concept 13, the Casimir effect, is a result of the temporalistic acceleration field in concept 1.
New recent observations (Benoit Famaey and colleagues – Strasbourg Observatory - G. Gentile et al. Nature, 461, 627-628, 2009) confirm the correlation between luminous matter and dark matter: In galaxies, "there is an inverse correlation between the central density of dark matter and that of visible matter. A high central density of visible matter implies that the density of dark matter at the center is low, and vice versa."

This observation is consistent with the temporalistic model. According to this model, the origin of dark matter lies in luminous sources (see Chapter X – Dark matter).

In addition, let us recall that the constant To, a quantum parameter, plays an important role in 4 quantum effects:

1) The elementary electric charge, e: h/bar × To.
The role played by To in the definition of electric charge, e, which is identical for all elementary particles but puzzling, can be explained by the absence in their definition of specific factors, such as mass, energy, spin, etc.

2) The constant of proportionality in the Josephson effect: 2 e / h, that is 2 e / h × 2μ, and, in angular frequency, 2 To

3) The constant of proportionality of the stopping potential in the photoelectric effect equals 1 / To.

4) In the temporalistic model, the fine structure constant turns out to be the ratio between the elementary electric charge e and the parameter G' (c / To).

CONCLUSION

An in-depth analysis of the most important concepts in the Big Bang model can only lead an impartial reader to one inescapable conclusion: contrary to the assertions of its supporters, the Big Bang model provides no evidence of its pertinence. A hypothesis (or an interpretation) can in no way be considered as evidence. The much vaunted example of the three pillars of the Big Bang model provides a blatant demonstration of this. Redshifts are facts and not evidence. The expansion of spacetime is only an interpretation and not evidence. Similarly, the cosmic microwave background, a cosmological observation, is interpreted as being fossil radiation. Yet again, this is an interpretation and not evidence. As for the third pillar, primordial nucleosynthesis, this is, once more, not evidence but a hypothesis about the possible formation of the most abundant chemical elements in the Universe. The same applies, as we have shown, to the other concepts mentioned in the previous paragraph.

The standard model of cosmology results from the erroneous interpretation of redshifts of distant galaxies. Since it started out from erroneous premises, it was understandable that, in order to develop, the Big Bang model found itself forced to fall back on a number of hypotheses that were inevitably wrong. These concepts are highly questionable and indeed flimsy from a scientific point of view, that is with regard to rigorous ananthropic concepts.

Unverifiable hypotheses lead to other even more unverifiable hypotheses (inflationary theories). The concepts in the Big Bang model, which are often highly speculative and 'unfalsifiable', are questioned and looked on with suspicion even by supporters of the Big Bang model (see inflationary theories - James Peebles).

Thus the Big Bang model, which has near-dogma status at present, is based on 11 hypotheses and 2 interpretations of important concepts which have not been validated, and provides strictly no evidence. In fact, it is a Ptolemaian-type model where hypothesis is piled up on hypothesis in an attempt to get round difficulties, without providing any evidence. In general, the difficulties and counter-examples are swept under the carpet and conveniently forgotten (for instance, the fact that the cosmic microwave background had been predicted by many other scientists long before Gamow's prediction, which was marred by errors, or the impossibility of explaining the origin of the primordial explosion because of its singularity, etc). It is in this respect that the Big Bang model is a delusion.

The alternative to the Big Bang model that we propose, the temporalistic model, is based fundamentally on the interpretation of redshifts. Whereas the Big Bang model interprets redshifts as being <u>spatial</u> phenomena, the temporalistic model considers them to be <u>temporal</u> phenomena. <u>The interpretation in the temporalistic model eliminates all the problems raised by the hypotheses and interpretations in the Big Bang model that we mentioned earlier.</u> Only two interpretations in the temporalistic model, the interpretation of redshifts as being temporal phenomena and the interpretation of the temporalistic accelerational field as being dark matter, need to be validated. The first of these is validated by its various consequences: the theoretical prediction, in 1962, of the value of the Hubble constant, H_o, verified in 2008 by NASA, and of the temporalistic constant T_o – Chapter VIII; the second is validated by the various pieces of evidence that we have set out: the correlation established between luminous matter and dark matter – Chapter X; validation of temporalistic gravitation by the agreement with the theoretical prediction of a large number of gravitational radii of stars, galaxies, etc – Chapter XII; Pioneer effect, Casimir effect, dark matter – Chapter X, etc.

A certain number of researchers have adopted the Big Bang model 'by default', since all the alternative models have been rejected. Such an attitude does not resolve the problem. Just because the alternative models may be wrong it does not necessarily follow that the standard cosmological model is right. The pertinence claimed for this model is, according to our analysis, a genuine delusion.

Like any other researcher, the author only believes in facts. Unlike the Big Bang model and its pseudo-evidence, the author proposes a certain number of genuine pieces of evidence (23), i.e. evidence that has been validated (rather than non-validated and frequently 'unfalsifiable' interpretations and hypotheses): 1) theoretical prediction, in 1962, of the value of the Hubble constant, Ho, and of the temporalistic constant To; 2) dark matter (origin and strength); 3) the Pioneer effect; 4) an alternative to MOND theory; 5) the Casimir effect; 6 to 17) masses and gravitational radii; 18 to 21) four quantum constants; 19) correlation between luminous matter and dark matter.

This evidence results from phenomena belonging to highly diverse areas of physics, including redshifts, the Hubble constant, gravitational radii of celestial objects, quantum constants, 'age of the Universe', dark matter, etc, which considerably strengthens the credibility of the temporalistic model.

It is worth noting that, according to Ockham's razor, i.e. the model that makes the least assumptions, the temporalistic model (1 hypothesis, 2 interpretations and 23 pieces of evidence) is totally pertinent, whereas the Big Bang model is not at all pertinent (11 hypotheses, 2 interpretations – no evidence).

The comparison between the Big Bang model and the temporalistic model enables us to point to those areas of research which seem to us to be fruitless, due to the initial erroneous nature of the standard Big Bang model, and to propose areas of research which we feel are likely to produce new, fruitful information about the structures and working of our Universe, together, naturally, with indisputable validations.

Fruitless research

1) The 'first' stars and galaxies
2) Inflationary theories
3) The singularity and the origin of the Big Bang
4) The horizon, flatness and critical density problems
5) Expansion and its acceleration – Dark energy
6) The cosmological constant Λ
7) 'Primordial' nucleosynthesis

Fruitful research

1) The fundamental test of the interpretation of redshifts: temporal effect or spatial effect. If this test could be carried out with accurate measurements it would make it possible to decide, unambiguously, between the Big Bang model and the temporalistic model.
2) The cosmic microwave background
3) The problem of the homogeneous and isotropic Universe
4) The Hubble constant, Ho
5) The quantum constant, To
6) The 'age' of the Universe
7) Luminous sources and dark matter
8) 'Temporalistic' gravitation – gravitational radii
9) The 'temporalistic' horizon
10) the Pioneer effect – the Casimir effect

And many other innovative avenues of research!

TESTS

Certain observational facts or tests should make it possible to choose unambiguously between the temporalistic model and the Big Bang model.

1) According to the Big Bang model, expansion begins beyond the Local Group of galaxies. According to the temporalistic model, redshift begins as soon as the photon is emitted. Would it perhaps be possible to undertake a statistical analysis of the speeds of stars at the boundary of our Local Group, with the aim of showing consistent redshift of radiation according to distance (or duration), even within the boundaries of the Local Group?

2) If it can be carried out, another observational test could make it possible to make a decisive choice between the spatially stationary Universe of the temporalistic model and the expanding Universe of the Big Bang model. If we compare the spectra of galaxies at, for instance, a fifty year interval, there are two possible outcomes. In the temporalistic Universe, the spectra of distant stationary galaxies, located 13 – 14 billion light years away, should not have varied. On the other hand, in an expanding Universe, the galaxies that are being carried away by expansion at relativistic speeds will be further away after 50 years, and will have speeds and redshifts that are different from those they had 50 years earlier. If it were possible to show the difference or lack of difference in the wavelengths of the spectra of

distant galaxies in two separate observations at 50-year intervals, the result would constitute a decisive test that would choose between the stationary temporalistic model and the cosmological Big Bang model and its expanding Universe.

3) It would be possible to contemplate two series of tests of the temporalistic model, one based on space, and the other on time. In the first category, the previous test would make it possible to choose between a stationary temporalistic Universe and an expanding Universe. Other tests could be based on the time coordinate. For instance, according to the temporalistic model, radiation traveling through space undergoes redshift. This redshift, which is caused by the existence of the temporalistic constant To, does not depend on the distance traveled through space but on the time that has elapsed. It would therefore be possible to contemplate a test that could detect the temporal or temporalistic redshift of radiation according to the elapsed time. An experiment such as the VIRGO project, where a laser beam travels along an optical path 150 km long could possibly be used to confirm or invalidate the temporalistic model. Other analogous tests could be thought up. According to the temporalistic model, a laser beam reflected back and forth between two mirrors during a certain time should undergo redshift. According to the spatial Big Bang model, the beam should show no redshift. A future lunar base could possibly used for this experiment. If this test could be carried out with accurate measurements it would make it possible to choose, unambiguously, between the Big Bang model and the temporalistic model.

4) In the area of gravitation, the temporalistic model proposes a finite range for gravitational fields, unlike other theories of gravitation. Chapter XII (Temporalistic gravitation) lists ten cases which confirm this proposition. If the infinitesimally small effects are measurable, it might be possible to contemplate verifying the finite range of gravitation in experiments similar to that of Etwöös torsion bars.

As can be seen, the temporalistic model proposes a number of possible tests of its refutability, in the Popperian sense.

Key words and key concepts in the temporalistic model

Anthropic and ananthropic concepts - temporalistic gravitation – the constant To, a quantum parameter, appears in 4 quantum effects: 1) the elementary electric charge, e: h-bar × To; 2) the constant of proportionality in the Josephson effect: 2 e / h, that is 2 e / h × 2μ, or, in angular frequency,

2 To; 3) the constant of proportionality of the stopping potential in the photoelectric effect equals $1 / T_o$; 4) the fine structure constant turns out to be the ratio between the elementary electric charge e and the parameter G' (c / T_o). - the temporalistic gravitational constant $G' = 6.582 \times 10^{-8}$ cm/sec² = acceleration of the gravity field = acceleration of dark matter; the temporalistic horizon, $T_o = 4.5546 \times 10^{17}$ s (approximately 14.43 billion years); the Hubble constant, $H_o = 67.71$ km/s/Mpc (these two last concepts were established theoretically in 1962); dark matter; Pioneer effect; MOND theory; Casimir effect; gravitational radius: $r = m^{1/2}$; ananthropic probabilistic model of the Universe; redshifts; cosmic microwave background; evolution of galaxies; tests of the temporalistic model.

OTHER RESEARCH BY THE AUTHOR

A PROBABILISTIC MODEL OF THE UNIVERSE:
www.site.voila.fr/probability
A PROBABILISTIC MODEL OF BIOLOGICAL EVOLUTION:
www.site.voila.fr/dinosaurs
A UNIVERSE WITH NO BIG BANG
www.site.voila.fr/nobigbang

PART EIGHT

Calculations: Chapter XV

Chapter VII: page 61

The theoretical prediction of the Hubble constant, Ho

Chapter VIII: page 70

Redshifts

In the expansion model, the redshift z at non-relativistic speeds due to the radial cosmological effect is given by the equation $z = v_r/c$. c is a speed that in a vacuum cannot be exceeded by any other physical speed. v_r is the radial velocity. c is a limiting constant. In the temporalistic model, the constant T_o is, similarly, a limiting constant for time periods. The redshift over short time periods is given by the equation $z = t / T_o$.

For non relativistic speeds, redshifts are given by the expression: $z = \lambda' - \lambda / \lambda = v_r / c$. (where z is the redshift, λ' the observed wavelength, λ the emitted wavelength, and v_r the radial velocity).

In calculating the redshift or 'recession effect', we did not take into account relativistic correction. At high, or more specifically, relativistic speeds, in other words close to the speed of light, redshift and the 'recession effect' are different, as can be observed in the spectra of distant quasars. The shift in wavelength can be in the region of several times its original value, and the 'recession effect' several times c.

At relativistic speeds, the relativistic relationship of the radial cosmological effect is given by the expression:

$\lambda' / \lambda = 1 + v_r/c / (1 - v_r^2/c^2)^{\frac{1}{2}}$

or $z = \lambda' - \lambda / \lambda = 1 + v_r/c / (1 - v_r^2/c^2)^{\frac{1}{2}} - 1$

The relativistic correction of redshifts and of the recession speeds of distant galaxies applies in the expanding Universe. This is due to the upper limit on the speed of light, a premise accepted in the expanding Universe model as well as in the temporalistic model, and the resulting slowing down of clocks. However, relativistic correction cannot play a role in the temporalistic Universe because it concerns light sources moving at relativistic speeds. In the temporalistic model, it is the radiation that varies, while the galaxies are stationary. Here, the 'recession effect' is an apparent effect and does not correspond to a cosmological effect at relativistic speeds. The relativistic redshift, at large distances or over large periods of time, nonetheless remains an experimental fact, and one which cannot be explained in the temporalistic model by a relativistic effect since the luminous sources are stationary. So how, then, are we to interpret it in the temporalistic model?

In the temporalistic model, speeds are replaced by times, and we obtain:

$\lambda' / \lambda = 1 + t_o / T_o / (1 - t^2 / T_o^2)^{\frac{1}{2}}$

or $z = \lambda' - \lambda / \lambda = 1 + t_o / T_o / (1 - t_o^2 / T_o^2)^{\frac{1}{2}} - 1$

where λ is the emitted wavelength and λ' the wavelength of the observed wavelength.

According to the temporalistic model, the redshift is therefore caused by the existence and effect of the temporalistic quantum constant $T_o = 4.55465 \times 10^{17}$ s (see Chapter XI – the ratio c / G). The 'pseudo-recession speed' of galaxies is only a 'recession effect' interpreted as a cosmological effect. The temporalistic constant To gives its theoretical value, which is precisely the one measured in observations of redshifts of distant galaxies.

The 'pseudo-recession speed' of galaxies at a distance of 1 Mpc can thus be calculated using the equation:

$v = Ho \times D = 2.997925 \times 10^{\wedge 10}$ cm/s $\times 10.287 \times 10^{\wedge 13}$ s / $4.55465 \times 10^{\wedge 17}$ s = 6.771×10^6 cm/s = 67.71 km/s/Mpc.

Estimated by Hubble in 1929 to be 500 km/s/Mpc (with an age of the Universe of 2 billion years), the 'pseudo-recession speed' of galaxies is today converging (after decades and more than 153 000 observations of redshifts by NASA) towards the value of 67.71km/s/Mpc established theoretically in 1962 by the author. <u>This theoretical value of Ho was obtained using purely physical arguments, independently of any astronomical data</u>, which bolsters its validity (see Chapter VIII).

Temporalistic redshifts, over temporalistic time periods, are similar to relativistic redshifts at relativistic speeds. The essential difference between the relativistic redshift and the temporalistic redshift stems from the origin of the redshift. On the one hand, this is a factor that is external to the radiation, the expansion of spacetime, and on the other, it is the quantum temporalistic effect, which is an integral part of the radiation.

The new explanation of the redshift z of distant galaxies put forward by the temporalistic model naturally has major cosmological implications.

The redshift z, or temporalistic effect, or 'recession effect' of galaxies according to their distance from the observer (or the travel time of the radiation) can be shown by:

$z = vr/c$ (in the cosmological effect) = t / To (in the temporalistic effect), where t = travel time of the photon (or distance / c) and To is the temporalistic constant. In the cosmological effect, the recession speed is given by $vr = z \times c$. For a redshift of 200 angströms for radiation of 4000 angströms, we get: $200 / 4000 \times 2.997925 \times 10^8$ m/s = 1.4989×10^7 m/s = 14 989 km/s.

For 1 second: 2.997925×10^8 m/s $\times 1$ s / $4.5546 \ 10^{17}$ s = 6.582×10^{-10} m/s = 6.582×10^{-8} cm/s.

For a duration corresponding to a distance of 1 Mpc: 2.997925×10^8 m/s $\times$ 10.287×10^{13} s / 4.5546×10^{17} s = 6.771×10^4 m/s = 67.71 km/s.

In the temporalistic model, $t = z \times To = 200 / 4000 \times 4.5546 \times 10^{17}$ s = 2.2773×10^{16} s and the recession effect $vr = c \times t / To = 2.997925 \times 10^8$ m/s $\times 2.2773 \times 10^{16}$ s / 4.5546×10^{17} s = 1.4989×10^7 m/s = 14 989 km/s.

If we apply to Hubble's law v (speed in km/s) = Ho (in km/s/Mpc) $\times$ d (distance in Mpc) the recession effect for 1 Mpc, we get Ho = v / d = 67.71 km/s / 3.084×10^{19} km (10.287×10^{13} s $\times 2.997925 \times 10^5$ km/s) = 2.195×10^{-18} s, that is, 1 / 4.5546×10^{17} s.

Since the constant To corresponds to an upper limit for duration in the same way as c is an upper limit for speed, the redshift is given by an expression different from $z = t / To$. Since the temporalistic constant plays the same limiting role with regard to time as the constant c does with regard to speed, the redshift for temporalistic durations (approaching 4.55456×10^{17} s) is therefore given by an expression similar to that of relativity, with speeds being replaced by times:

Interpreted as a cosmological effect, the redshift is considered to be caused by the recession of galaxies and its value depends on the Hubble constant Ho according to the equation: $v = Ho \times D$ (1)

where v is recession speed, Ho the Hubble constant and D the distance of the galaxy.

We saw earlier that redshift in the cosmological effect $z = vr / c$ is interpreted in the temporalistic model by $z = t / To$, where z is redshift, vr radial velocity, c the speed of light, t the travel time of the photon (or distance / c) and To the temporalistic constant, from which we get:

$z = vr / c = t / To$ and $vr = ct / To$ (2)

Applying equation (2) to equation (1), we get:

$vr = Ho \times D = ct / To$, and since $D = ct$, we obtain $vr = Ho \times ct = ct / To$

From this we get:

$Ho = 1 / To = 1 / 4.55465 \times 10^{17}$ s

According to the temporalistic model, as soon as the photon is emitted, the existence of the temporalistic constant To becomes apparent by a reddening (redshift) of its wavelength without any external intervention. To explain the redshift of distant galaxies, the temporalistic model does not therefore need the various expanding Universe (FLRW) cosmological models.

The interpretation by the Big Bang model of the redshift of distant galaxies as being a cosmological effect caused by the expansion of spacetime, is also refuted by the temporalistic model. The cosmological effect $z = vr / c$ is interpreted in the temporalistic model by $z = t / To$, where z is the redshift, vr the apparent radial velocity, c the speed of light, t the travel time of the photon (or distance / c), and To the temporalistic constant.

Whereas, in the Big Bang model, expansion only begins beyond the local system of galaxies, in the temporalistic model, redshift (or recession effect) takes place as soon as the photon is emitted.

If we apply to Hubble's law v (speed in km/s) = Ho (in km/s/Mpc) × d (distance in Mpc) the recession effect for 1 Mpc, we get Ho = v / d = 67.71 km/s / 3.084×10^{19} km (10.287×10^{13} s × 2.997925×10^5 km/s) = 2.195×10^{-18} s, that is, 1 / 4.5546×10^{17} s.

The value of the temporalistic effect or 'recession effect' at 1 Mpc = 67.71 km/s and that of Ho = 1 / 4.5546×10^{17} s (approximately 14.43 billion years) were established <u>theoretically</u> by the author in <u>1962.</u> The latest data provided by WMAP 5 (Table 7 – Cosmological Parameter Summary – 2008) gives a value for Ho = 71.9 (+2.6 – 2.7) km/s/Mpc and to = 13.69 (± 0.13) billion years.

Comparing the observational value and the theoretical value for Ho, 69.2 km/s/Mpc (71.9 – 2.7) for the former and 67.71 km/s/Mpc for the latter, i.e. a difference of 2.16%, the difference is negligible if we take into account the range of uncertainty in the WMAP 5 data: from 3.2% (+2,6) to 3.75% (-2,7). We should add that the value of Ho provided by WMAP 5 was obtained after 80 years of research and corrections, of which 69.2 km/s/Mpc is the most recent but certainly not the final result, whereas the theoretical value proposed by the author <u>as long ago as 1962</u> has not changed since then.

Chapter XI: page 116

The temporalistic model – The concept of time and the constant To – The temporalistic hypothesis - Searching for the constant To – The ratio c / G – The quantum constant G': 4 quantum effects

In Newtonian mechanics, we have (in the CGS system): $To = c/G$, that is, 2.99792×10^{10} cm/s / 6.67×10^{-8} cm^3/g-s^2 = 4.494×10^{17} s g/cm^2. (1)

The value of To would be 4.494×10^{17} s if the ratio g/cm^2 was approximately equal to one.

Or

$To = (c / G) (An / Mn)$

Where

$c = 2.99792 \times 10^{10}$ cm/s
$G = 6.67 \times 10^{-8}$ cm^3/g s^2
$Mn = 1.67 \times 10^{-24}$ g (mass of the neutron or proton)
An = approximately 1.67 barn = 1.67×10^{-24} cm^2 (scattering cross section of a proton or neutron)

The value of To would be 4.494×10^{17} s if the ratio An / Mn (cm² / g) was approximately equal to one.

Gravitational interaction concerns the effect of masses (and energy) on other masses or on the metric field. It operates on the level of particles and more specifically on the level of atoms and molecules (protons, neutrons, electrons). It does not operate on the subatomic level of the strong nuclear force (quarks and gluons), and does not therefore concern quantum chromodynamics.

In the final analysis, gravitation, the interaction between masses (and energy), therefore concerns nucleons (the protons and neutrons of which atoms are made), and, on larger scales, astronomical masses (planets and satellites, stars, galaxies, etc).

Hydrogen and helium are the most abundant elements in the Universe:

Hydrogen makes up about 94% of the Sun in terms of the number of atoms and 73% by mass, while helium makes up 5.9% and 25% respectively. Hydrogen makes up about 85% of the Universe in terms of the number of atoms and 66% by mass, while helium makes up 13% and 31% respectively.

The cross section of nucleons (protons and neutrons) plays a fundamental role in the phenomenon of gravitation.

The barn, 10^{-24} cm², represents a very small area. This is the order of magnitude of the cross section of a large atomic nucleus. The cross section has nothing to do with the 'geometric properties' of nuclei and has no specific relationship to its size. It is linked to the energy of incident particles. In general, the lower the energy, the larger is the cross section. We should not forget that, according to quantum mechanics, the scattering of particles is the result of interactions between waves and other waves.

The cross sections of the reactions of the proton and the neutron are very similar, once the effects of the proton's electric charge have been taken away.

The limit of the cross section of coherent scattering of the neutron by the isotope ^{1}H (whose abundance is 99.985%) is 1.7583 barn. For the isotope ^{4}He (whose abundance is 99.99986%), it is 1.34 barn. (NIST Center for Neutron Research - http://www.ncnr.nist.gov/resources/n-lengths/list.html). The cross section of coherent scattering of the neutron by the isotope ^{1}H, for a wavelength of 1 angstrom, 1.76 barn, is confirmed on the website (http://www.11b.cea.fr/pedagogie/absortrayonsx/absortrayonsx.html).

An experiment carried out by a team from GSI (Darmstadt, Germany) on targets of deuterium and hydrogen at incident energies of 800 MeV to 1 GeV per nucleon, using gold, uranium and lead projectiles, enabled them to obtain a total cross section of 1765 mb (1.765 barn) with a precision of under 5%, "which agrees with measurements by other teams". (http://www.google.fr/search?q=cache:bhNWEprIfqoC:wwwcenbg.in2p3.fr/extra/Noy-ex...)

Another experiment carried out at GSI (Darmstadt, Germany) of interaction on a ^{8}B proton halo target gives a cross section of around 1.5 barn with an incident energy of 20 MeV/nucleon (http://wwwcenbg.in2p3.fr/extra/Noy-exotique/7Be.html).

The mean cross section of nucleons can be estimated to be equal to or close to 1.7 barn. The mass / cross section ratio of the proton and the neutron is thus roughly equal to one: 1.67×10^{-24} g/1.7×10^{-24} cm² ~ 1 g/cm², so g/cm² ~ 1.

In the temporalistic model, equation (1) becomes $c / G = To$, that is 2.99792×10^{10} cm/s / 6.67×10^{-8} cm^3/g s^2 = 4.494×10^{17} s g/cm^2, and with g/cm$^2 \sim 1$, $c / G = To = 4.494 \times 10^{17}$ s, which is approximately 14.24 billion years.

In the temporalistic model, the Newtonian gravitational constant, 6.67×10^{-8} cm^3/g s^2, is thus interpreted, with g/cm² ~1, as being the temporalistic gravitational constant G':

$G' = 6.67 \times 10^{-8}$ cm/s² (2)

In the temporalistic model, equation (1) c / G becomes:

$c / G' = To$, that is, 2.99792×10^{10} cm/s / 6.67×10^{-8} cm/s² = 4.494×10^{17} s, which is approximately 14.24 billion years.

Quantum physics tells us that different atomic nuclei have binding energies of varying sizes (Aston's packing fraction), and as a result, have a mass defect. The binding energy per nucleon for nuclei containing between 30 and 120 nucleons is over 8.5 MeV. It is around 9 MeV for nuclei with a mass number in the region of 56 (Fe). It is therefore necessary to adjust the 'macroscopic' value of G' in relation to the masses of quantum particles without nuclear binding energy, electrons, nucleons, etc. To a first approximation, the 'quantum' value of G' is therefore 6.60×10^{-8} cm/s² and To = 4.5423×10^{17} s, which is approximately 14.4 billion years.

Later on we will be able to refine this value of To even further, using purely quantum constants that are more precise. The value of To = 4.55465×10^{17} s was established by the author in 1962. (See Chapter XI – The ratio c / G page 120)

Chapter XI: Four quantum effects page 111

The constant To, a quantum constant, appears in 4 quantum effects:

1) The elementary electric charge, e: <u>h-bar × To</u>
2) The constant of proportionality in the Josephson effect: 2 e / h, that is 2 e / h × 2μ, and, in angular frequency, <u>2 To</u>
3) The constant of proportionality of the stopping potential in the photoelectric effect equals <u>1 / To</u>.

4) In the temporalistic model, the fine structure constant turns out to be the ratio between the elementary electric charge e and the parameter G' (c / To), that is, e / c × To

The elementary electric charge e

In Chapter XI, we showed that the temporalistic model postulated the existence of the constant To in the physics of the photon. The redshift of galaxies is no longer interpreted as being a cosmological effect but is rather considered to be an intrinsic quantum property of the photon.

According to quantum physics, there is no fundamental difference between the photon and the electron. The emission of photons from atoms is caused by transitions in the energy levels of photoelectrons. The photoelectric effect demonstrates the transfer of energy of incident photons to the electrons of illuminated metal and the disappearance of the incident photons. Particles of matter (electrons) or of energy (photons) appear and disappear, replacing each other, but the energies and momentum are conserved. A positron and an electron that collide can annihilate each other, forming two γ rays. To a first approximation, the photon can be considered to be a kinetic particle of energy moving through space and the electron the corresponding particle of matter, relatively static and rotating. If this analysis is correct, the physics of the electron, like that of the photon, must incorporate the temporalistic quantum constant To. In Chapter XI, we saw how the physics of the photon is affected by To. Here we shall examine how the physics of the electron incorporates the constant To.

In the standard particle model, an important property of particles is their spin, whose unit is h-bar. Spin can be defined as the intrinsic angular momentum of particles. In addition, another important quantum property of particles is their electric charge e. The value of this charge is the same for all free charged 'elementary particles' (whereas quarks and antiquarks, which have fractional charge, are confined): this is the elementary electric charge ± e, $4.8032068 \times 10^{-10}$ esu cgs units or 1.60218×10^{-19} coulomb in SI MKSA units.

In Chapter XI, we saw that the constant To can be considered, similarly to the constant c, to be a quantitative and limiting constant for quantum phenomena. Just as c is a limiting parameter for physical speeds and a quantitative parameter for the energy of matter at rest $E = mc^2$, could To not be a limiting and quantitative temporalistic constant for the motion of particles,   or more specifically, for the intrinsic angular momentum of particles, in other words for their spin h-bar? From this perspective, we can set down the equation: total spin = h-bar (spin or angular momentum) × To (time) = h-bar × To (total angular momentum). As dimensions, $ML^2T^{-1} \times T = ML^2$ (moment of inertia). In CGS numerical values: $1.05457266 \times 10^{-27}$ erg s × 4.5423×10^{17} s = 4.790185×10^{-10} erg s². The numerical value of the angular momentum or total spin is remarkably close to the numerical value of e = $4.8032068 \times 10^{-10}$ esu cgs units. Is this just a simple coincidence? We do not believe so. The numerical value of the total spin, h-bar × To, so close to that of e, would probably be identical if the numerical value of G' (6.60 × 10^{-8} cm/s²) had been experimentally established on the basis of quantum measurements rather than on macroscopic measurements (Cavendish's torsion balance experiment).

The temporalistic hypothesis of the quantum constant To leads us to propose that the elementary electric charge e should be considered as being the total action or total spin (total intrinsic angular momentum) of particles, that is, e = h-bar × To, and To = e / h x 2 µ. From this perspective, we can assign to To the definitive 'quantum' numerical value of e / h x 2 µ = 4.55465×10^{17} s, or 14.43 billion years, and to G' the 'quantum' value c / To = 2.99792×10^{10} / 4.55465×10^{17} = 6.58210×10^{-8} cm/s², which is very close to its macroscopic value (6.60×10^{-8} cm/s²).

Let us now examine the numerical and dimensional problems raised by the temporalistic definition of the elementary electric charge e.
a) In the CGS esu system: in the temporalistic model, the numerical value of e is given by e = h-bar × To = $1.05457266 \times 10^{-27}$ erg s × 4.55465×10^{17} s = $4.8032068 \times 10^{-10}$ erg s². In the CGS esu system, e = $4.8032068 \times 10^{-10}$ cgs esu. The dimension of e is specific. In the temporalistic model, the dimension of e (h-bar × To) is $ML^2T^{-1} \times T = ML^2$ (ergs s²). This is the dimension of a moment of inertia. In the temporalistic model, the electric charge e does not have a specific CGS dimension and can therefore be incorporated into the three dimensions L, M and T. Neither does it require the existence of a dimension of electric charge in esu.

b) In the MKSA system let us compare the values of h and e in the CGS esu and MKSA systems.

h (cgs esu) / h (MKSA) = $1.05457266 \times 10^{-27}$ erg s / $1.05457266 \times 10^{-34}$ J s = 10^7

e (cgs esu) / e (MKSA) = $4.8032068 \times 10^{-10}$ esu / 1.6019×10^{-19} C = 2.99792×10^9

Hence the ratio h (cgs) / h (MKSA) / the ratio e (cgs esu) / e (MKSA) = 10^7 / $2.99792 \times 10^9 = 1 / 2.99792 \times 10^{-2}$

If we set down, in the MKSA system, e = h-bar × To, we obtain e = $1.05457266 \times 10^{-34}$ J s × 4.55465×10^{17} s = $4.8032068 \times 10^{-17}$ J s^2, consistent with e = $4.8032068 \times 10^{-10}$ erg s^2 but different from 1.6019×10^{-19} C. To obtain this value, we have to take into account the inconsistency of the relationships between h in the CGS esu and MKSA systems on the one hand and e in the same systems, that is, e in the MKSA system = h-bar × To = $1.05457266 \times 10^{-34}$ J s × 4.55465×10^{17} s × $1 / 2.99792 \times 10^{-2}$ = 1.6019×10^{-19} J s^2. In the MKSA system, the temporalistic dimension of e is, as in the CGS esu system, defined by e = h-bar × To, that is, $ML^2T^{-1} \times T = ML^2$ (moment of inertia). It no longer requires the existence of a specific dimension of e in coulombs (or amperes).

The temporalistic dimension of e (ML^2), as well as its numerical value is justified by its consistency in quantum phenomena. We can note here that the identical value, for all 'elementary particles', of the elementary electric charge e, which is observed in quantum electrodynamics but not explained, is easily explained in the temporalistic model. Its definition (<u>the product of two universal constants, h-bar and To</u>) does not involve the specific properties of particles (mass number, energy, baryon or lepton number, etc). <u>The identical electric charge of two completely dissimilar particles such as the positron and the electron is thus explained.</u> The fractional charge of quarks is not operational since they are confined. In this connection, we can call upon a quantum principle: the conservation of total angular momentum in quantum systems. It is likely that a similar principle applies to the fractional charge of quarks, since the elementary electric charge e appears as the total angular momentum of particles.

The temporalistic proposition for the elementary electric charge e = h-bar × To allows us to set down e / h - bar = To or h - bar / e = 1 / To. The constant To leads us directly into the heart of quantum physics, where this relationship between e and h appears in the Josephson effect, the photoelectric effect and, indirectly, the fine structure constant.

The Josephson effect

The Josephson effect arises when a current of electrons passes between two strips of superconducting material (lead, aluminium, niobium, etc) separated by an insulating barrier. When such superconducting materials are cooled to a very low temperature (a few kelvins), free electrons form Cooper pairs. Quantum mechanics explains this fact by saying that all the pairs condense to the same quantum state described by a single macroscopic wave function. The strength of the current produced only depends on the phase difference between the wave function on the one hand and the barrier on the other. It varies as the sine of the phase difference, a phase difference whose time derivative is itself proportional to the voltage on either side of the barrier. The proportionality constant is proportional to e / h. For a constant voltage V, measured at the terminals of a Josephson junction, a sinusoidal current of frequency ν = 2 e / h × V is seen to flow. The Josephson effect makes it possible to relate, via the two universal constants e and h, voltage and frequency. By international agreement, the value of the ratio frequency / voltage proportional to 2e / h is defined as 483 594 Ghz/V.

Let us set down the dimensional equation of the proportionality constant 2e / h: $Q/ML^2T^{-1} = C/J$ s = esu/erg s or frequency / voltage = $T^{-1}/ML^2T^{-2}Q^{-1}$ = $1/ML^2T^{-1}Q^{-1} = Q/ML^2T^{-1} = C/J$ s = ues/erg s whence ν = 2 e / h × V = 2 × C/J s × J/C = 2 × Q/ML^2T^{-1} × $ML^2T^{-2}Q^{-1}$ = 2 T^{-1}.

In numerical values, the proportionality constant equals $2 \times 1.60217 \times 10^{-19}$ C / 6.626075×10^{-34} J s = $2 \times 2.41797 \times 10^{14}$ = 4.83594×10^{14} C/J s or 483 594 Ghz/V. We can calculate the angular frequency ω = 2μ ν, that is, 2μ × 2.41797×10^{14} Hz = 1.519259×10^{15} Hz and the corresponding proportionality constant $2 \times 1.519259 \times 10^{15}$ Hz/V.

We shall now introduce the dimensions and numerical values of the temporalistic model, in the CGS system. The dimensional equation of the proportionality constant 2 e / h = ML^2 / ML^2T^{-1} = T = frequency / voltage = $T^{-1} / ML^2T^{-2}Q^{-1} = T^{-1} / T^{-2}$ = T. In numerical values, 2 e / h × 2μ (in angular frequency) = 2 × h-bar × To / h × 2μ = 2 × To, that is, $2 \times 4.8032068 \times 10^{-10}$ erg s² / 6.626075×10^{-27} erg s × 2μ = $2 \times 4.5546 \times 10^{17}$ s.

If the temporalistic interpretation is correct, it should be consistent with the quantum interpretation of the Josephson effect. Let us see if it is. In the temporalistic model, the proportionality constant 2 e / h has the dimension of time: To = T and its value is $2 \times 4.5546 \times 10^{17}$ s / 2μ in cgs esu units. In quantum theory, the proportionality factor has the dimension of frequency/volt and its value is $2 \times 1.51925 \times 10^{15}$ Hz/V. We know that in the SI MKSA system, the electric potential of one electrostatic unit is 299.792 volts. It is therefore the same thing if we give the proportionality constant the value $2 \times 1.51925 \times 10^{15}$ Hz/V or $2 \times 4.5546 \times 10^{17}$ Hz (1.519259×10^{15} ×

299.792) / 299.792 volts (1 esu). In dimensions, the proportionality constant is, as we saw, Hz / V = T^{-1} / $ML^2T^{-2}Q^{-1}$ = T^{-1} / T^{-2} = T, consistent with the dimension of the temporalistic proportionality constant To (T).

The proportionality constant of the Josephson effect 2 e / h, that is, 2 e / h × 2µ in angular frequency, is therefore equal to 2 To, which shows the presence of the temporalistic constant in this quantum effect.

The photoelectric effect

After the discovery of the Planck constant, h, Einstein hypothesized the corpuscular nature of the photon and set out his famous equation for the photoelectric effect E kin = hv − W, where W is the work function, i.e. the energy required to withdraw an electron from the material.

Later on, experiments by Millikan established a linear relationship between the stopping potential Vo and the frequency of the incident light Vo = h / e × v - We.

If the stopping potential Vo is plotted on a graph against the frequency v, we obtain a straight line equal to H / e, and, ignoring the work function of the material, the stopping potential of the photoelectric emission will be proportional to the constant h / e: Vo = h / e × v or Vo / v = h / e. Let us put in the numerical values: h / e = 6.626075 × 10^{-34} J s / 1.602177 × 10^{-19} C = 4.1357 × 10^{-15} V s. The dimensional equation of Vo / v gives $ML^2T^{-2}Q^{-1}$ / T^{-1} = $ML^2T^{-1}Q^{-1}$ = ML^2T^{-1} / Q. The proportionality constant of the stopping potential is therefore 4.1357 × 10^{-15} V s and the stopping potential Vo = 4.1357 × 10^{-15} V s × v or, by angular frequency ω = 2 µ v = 4.1357 × 10^{-15} / 2µ × v = 6.582 × 10^{-16} V s × v.

The photoelectric effect has similarities with the Josephson effect. In both quantum effects, an electric current is produced. In the Josephson effect, an electric current is produced across an insulating barrier, under certain conditions. There exists a proportionality constant between the frequency of the current produced and the voltage at the terminals of the Josephson junction. This proportionality constant, in CGS units, is e / h, that is, the temporalistic constant To / 2µ. In similar fashion, in the photoelectric effect there is a proportionality constant between the stopping potential of the electric current produced by the photoelectrons and the frequency of the incident radiation. This proportionality constant is h / e, i.e. in the temporalistic model, 2 µ / To, the reciprocal of the temporalistic constant.

Let us put in the temporalistic dimensions and numerical values.

The dimensional equation of the proportionality constant of the stopping potential gives h / e = voltage / frequency, that is, h / e = ML^2T^{-1} / ML^2 = T^{-1} or voltage / frequency = $ML^2T^{-2}Q^{-1}$ / T^{-1} = $ML^2T^{-1}Q^{-1}$ = T^{-1}.

In the CGS system, in numerical values we get h / e = h / h-bar × To = 2μ / To; using the angular frequency $\omega = 2\mu \nu$, we get h / h-bar × To 2μ = 1 / To, that is, 6.626075 × 10^{-27} erg s / 4.8032068 × 10^{-10} erg s^2 × 6.2832 = 2.1955 ×10^{-18} s.

If it is correct, the temporalistic model should converge with the quantum interpretation of the photoelectric effect. The dimensional equation of the proportionality constant of the stopping potential is, in the temporalistic model, T^{-1} (1 / To). In quantum theory, it is h / e, that is, $ML^2T^{-1}Q^{-1}$ (volt seconds), that is, converted to temporalistic dimensions, ML^2 (e) $T^{-1}Q^{-1}$ (-e) = T^{-1}.

In numerical values, in the temporalistic model, 1 / To = 1 / 4.5546 × 10^{17} s = 2.1955 × 10^{-18} s. In quantum theory, in SI, h / e = 6.626075 × 10^{-34} J s / 1.602177 × 10^{-19} C = 4.1357 × 10^{-15} V s, or by angular frequency, 6.582 × 10^{-16} V s. Let us introduce the potential in esu, that is, 299.7925 volts per esu. The proportionality factor is therefore 6.582 × 10^{-16} V s × 1/299.7925 V = 2.1955 × 10^{-18} s.

We can check the accuracy of this value by calculating the stopping potential of blue light with a frequency in the region of 7 × 10^{14} Hz: 2μ × 7 × 10^{14} s × 2.1955 × 10^{-18} s × 299.7925 V = 2.89 V, which is indeed the order of magnitude of the required stopping potential.

The proportionality constant of the stopping potential of the photoelectric effect is equal to 1 / To and the presence of the temporalistic constant is again found in this quantum effect.

The fine structure constant, α

The fine structure constant is one of the fundamental constants of Nature. It plays a major role in quantum electrodynamics. Let us briefly recall its essential characteristics. α is the coupling constant that describes the coupling of any elementary particle carrying an electric charge e with the electromagnetic field. The fine structure constant establishes the ratio of the electrostatic coupling energy between an electrical particle and the electric field on the one hand, and its rest mass energy on the other: $\alpha = e^2 /$ (h /mc) / $mc^2 = e^2$ / h bar $\times$ c = 7.2992 $\times$ 10^{-3} = 1 / 137.036, h / mc being the Compton wavelength of the electrical particle and mc^2 its rest mass energy.

The fine structure constant α also plays an important role in Feynman diagrams relating to electron-electron scattering processes. The contribution of each diagram to the scattering process rate is proportional to a particular power of the factor 1 / 137 (of the fine structure constant α), that is, $(1 / 137)^n$, where n can be 1, 2, 3, etc.

When we consider the fine structure constant α within the framework of the temporalistic model, we reach some interesting results. Let us apply the temporalistic constants e = h-bar $\times$ To and G' = c / To. We get $\alpha = e^2$ / h-bar $\times$ c = e / c $\times$ To.

1) In the CGS esu system: let us apply the numerical values. In quantum theory, $\alpha = e^2$ / h-bar $\times$ c = 4.8032068 $\times$ 10^{-10} $\times$ 4.8032068 $\times$ 10^{-10} / 1.054572 $\times$ 10^{-27} $\times$ 2.997925 $\times$ 10^{10} = <u>7.2974 $\times$ 10^{-3}</u>; in the temporalistic model, e / G' = 4.8032068 $\times$ 10^{-10} / 6.582 $\times$ 10^{-8} = <u>7.2974 $\times$ 10^{-3}</u>.

In dimensions: in quantum theory: e^2 / h c = $ML^3T^{-2}Q^{-2}$ $\times$ Q^2 / ML^2T^{-1} $\times$ LT^{-1} = ML^3T^{-2} / ML^3T^{-2} = <u>Dimensionless number</u> (whence e^2 = ML^3T^{-2}).

In the temporalistic model, e^2 = ML^3T^{-2} whence e / G' = e^2/e / G' = ML^3T^{-2}/ML^2 / LT^{-2} = LT^{-2} / LT^{-2} = <u>Dimensionless number</u>.
2) In SI MKSA: In numerical values: In quantum theory: e^2 / h c = 8.987 $\times$ 10^9 (constant K in a vacuum for SI) $\times$ 1.602 $\times$ 10^{-19} $\times$ 1.602 $\times$ 10^{-19} / 1.054 $\times$ 10^{-34} $\times$ 2.997925 $\times$ 10^8 = 2.306 $\times$ 10^{-28} / 3.16 $\times$ 10^{-26} = <u>7.2974 $\times$ 10^{-3}</u>.

In the temporalistic model: e / G' = e^2/e / G' = 2.306 $\times$ 10^{-28} / 1.602 $\times$ 10^{-19} / 6.582 $\times$ 10^{-8} = 2.1877.

Taking into account the inconsistencies between the CGS and MKSA systems: =
2.997925 $\times$ 10^2, e / G' = 2.1877 $\times$ 1/299.7925 = <u>7.2974 $\times$ 10^{-3}</u>.

In dimensions: in quantum theory: e^2 / h c = ML^3T^{-2} / ML^2T^{-1} $\times$ LT^{-1} = ML^3T^{-2} / ML^3T^{-2} = <u>Dimensionless number.</u>

In the temporalistic model, e / G' = e^2/e / G' = ML^3T^{-2} / ML^2 / LT^{-2} = LT^{-2} / LT^{-2} = <u>Dimensionless number</u>

We can see that the temporalistic constant To turns up in the definition of the fine structure constant α aince $\alpha = e^2 / (h/mc) / mc^2 = e^2 / h\ c = e/c \times$ To or e / G'. In quantum mechanics, α is interpreted as the coupling constant of electromagnetic interactions, or the ratio between the electromagnetic energy and the rest mass energy of any 'elementary' electrical particle. In the temporalistic model, the fine structure constant appears as the ratio between the elementary electric charge e and the parameter G' (c / To), that is, e / c × To.

Like any other researcher, the author only believes in facts. Unlike the Big Bang model and its pseudo-evidence, the author proposes a certain number of genuine pieces of evidence (23), i.e. evidence that has been validated (rather than non-validated and frequently 'unfalsifiable' interpretations and hypotheses): 1) theoretical prediction, in 1962, of the value of the Hubble constant, Ho, and of the temporalistic constant To; 2) dark matter (origin and strength); 3) the Pioneer effect; 4) an alternative to MOND theory; 5) the Casimir effect; 6 to 17 masses and gravitational radii); 15 to 18) four quantum constants; 19) correlation between luminous matter and dark matter; 20) The new measurement of the Hubble constant is 67.0 ± 3.2 km/s/Mpc", i.e. within 1% of the temporalistic value of Ho Florian Beutler et al. ICRAR (International Centre for Radio Astronomy Research) – UWA (University of Western Australia) 25 July 2011.

THE 23 PIECES OF EVIDENCE FOR THE TEMPORALISTIC MODEL

1) The author established, in a strictly theoretical manner, in 1962, a value for the Hubble constant, Ho, of 67.71 km/s/Mpc, validated by NASA with a value of 71.9 (WMAP 5 – Table 7 – Cosmological Parameter Summary – 2008) with an uncertainty of 3.2 % (+2 .6) to 3.75 % (-2.7), that is, 69.7 km/s/Mpc, which is a difference of 2 km/s/Mpc, within the range of uncertainties. (One piece)

2) "The new measurement of the Hubble constant is 67.0 ± 3.2 km/s/Mpc", i.e. within 1% of the temporalistic value of Ho. (one piece)

Florian Beutler et al. ICRAR (International Centre for Radio Astronomy Research) – UWA (University of Western Australia) 25 July 2011. one piece)

3) The temporalistic constant To = $1/H_o$ = 4.5546×10^{17} s - about 14.43 billion years. This value plays a role in several areas: deceleration of Pioneer 10, MOND theory, quantum effects, etc. one piece)

4) Dark matter (origin and value): dark matter originates in the luminous flux of radiation from stars which, as it travels through space, loses energy. This phenomenon gives rise to an acceleration field or graviton field with a value of 6.582×10^{-8} cm/s^2. This is also the value of the temporalistic gravitational constant G' = 6.582×10^{-8} cm/s^2.
The close relationship between luminous matter and dark matter is shown by new observations by Benoit Faley et al – Strasbourg Observatory – G. Gentile et al. Nature 461, 627 – 628 2009
See also Chapter X. (one piece)

5) Pioneer 10 anomaly – the order of magnitude of this anomaly is that of c × Ho (Hubble constant), that is, c / To temporalistic constant) = G' temporalistic gravitational constant, 6.582×10^{-8} cm/s^2. One experimental measurement validates the temporalistic model. By September 1998, the slowing down of Pioneer 10 had caused it be some 400 000 km closer to the Sun compared to its predicted path. The radial journey of Pioneer 1, which began in 1973 – 1974, had therefore lasted some 24.5 years, i.e. 7.73×10^8 s. During this period, the deceleration, with an acceleration constant of 6.582×10^{-8} cm/s^2, equalled 6.582×10^{-8} cm/s$^2 \times 7.73 \times 10^8$ s × 7.73×10^8 s =3.93293 × 10^{10} cm = 393 293 km. See Chapter X (one piece)

6) The MOND theory proposes that when the acceleration inferred from the Newtonian acceleration constant Gn is less than a°, that is Gn < a°, Newtonian theory does not apply, the parameter a° being comparable in value to c × Ho.
According to the temporalistic model where Ho = 1 /To, a° ~ c × Ho = c / To, which is 6.582×10^{-8} cm/s^2. MOND theory is proposed as an alternative to dark matter. The temporalistic model does not deny the existence of dark matter. When the acceleration caused by a mass is smaller than G', the Newtonian model no longer applies in MOND theory. In the temporalistic model, Newtonian theory no longer applies for an acceleration smaller than G', as in MOND theory, but this is due to the finite gravitational radius of masses and to the temporalistic universal acceleration field G'. (See Chapter XI).

The Big Bang model does not support MOND theory. (one piece)

7) The Casimir effect The temporalistic model proposes an alternative to the quantum explanation. The isotropic acceleration field of value G' created by the energy loss of photons is disturbed by the presence of the two metal plates. The result is that there is a smaller acceleration force between the two plates than on the outside of the plates, causing them to move closer to each other. It would be interesting to calculate whether this temporalistic effect is quantitatively confirmed. (one piece)

8) The constant To, a quantum parameter, plays an important role in 4 quantum effects:

1) The elementary electric charge, e: h-bar × To
2) The constant of proportionality in the Josephson effect: 2 e / h, that is 2 e / h × 2μ, and, in angular frequency, 2 To
3) The constant of proportionality of the stopping potential in the photoelectric effect equals 1 / To.
4) In the temporalistic model, the fine structure constant turns out to be the ratio between the elementary electric charge e and the parameter G' (c / To), that is, e / c × To (4 pieces)

9) The validation of the temporalistic model of gravitation is provided by many observational verifications (Chapter XII: Masses and gravitational radii), with 12 pieces of evidence for the range of the gravitational radius of masses, shown by the temporalistic model and given by the expression $r = m^{1/2}$ (r in cm = gravitational radius, m in g = mass). (12 pieces).

© 2012, Salomon Borensztejn
Edition : BoD - Books on Demand, 12/14 rond-point des Champs Elysées,
75008 Paris
Impression : Books on Demand, Allemagne
ISBN : 9782322023288
Dépôt légal : juin 2012

www.ingramcontent.com/pod-product-compliance
Lightning Source LLC
Chambersburg PA
CBHW050209230526
45470CB00001B/311